KU-319-613

The Hedgerow Handbook

Dedication

This book is for everyone who understands the simplicity of living in the moment, and who can appreciate a nice hedge when they see one.

For Mister Fitz, Sir Fitzy and Miss Fitz.

Disclaimer

The author and publisher disclaim, as far as the law allows, any liability arising directly or indirectly from the use, or misuse, of the information contained in this book. The information in this book has been compiled by way of general guidance in relation to the specific subjects addressed, but it is not a substitute and not to be relied on for medical, healthcare, pharmaceutical or other professional advice on specific circumstances and in specific locations.

So far as the author is aware the information given is correct and up to date as at March 2012. Practice, laws and regulations all change, and the reader should obtain up to date professional advice on any such issues.

Cynddylan on a Tractor

Ah, you should see Cynddylan on a
*   tractor.*
Gone the old look that yoked him to the
*   soil,*
He's a new man now, part of the
*   machine,*
His nerves of metal and his blood oil
The clutch curses, but the gears obey
His least bidding, and lo, he's away
Out of the farmyard, scattering hens.
Riding to work now as a great man
*   should,*
He is the knight at arms breaking the
*   fields'*
Mirror of silence, emptying the wood
Of foxes and squirrels and bright jays.
The sun comes over the tall trees
Kindling all the hedges, but not for him
Who runs his engine on a different fuel.
And all the birds are singing, bills wide
*   in vain,*
As Cynddylan passes proudly up the
*   lane.*

R. S. THOMAS, 1950

Contents

Introduction
The Great British Hedgerow

A bird's eye view

Imagine you're a tourist flying into Britain for the very first time. As the aircraft begins its descent and the clouds disperse you see the sea and the rugged coastline. Then, as the plane flies even lower and the land comes more clearly into focus, you see something you've never seen before: a seemingly random mosaic of lush fields in different shades of green, some acid yellows, some ochre, all bordered by (mostly) straight lines of trees and hedges, and perhaps the ribbon of a river glinting in the sun as it twists towards the horizon.

If there's one distinctive feature of the British countryside, it has to be the hedgerow. In fact, it's such an established part of the structure of our landscape that sometimes we hardly even notice it. But, without these hedgerows, the landscape would look very different and that traveller would see a very different picture …

A brief history of the hedgerow

It's estimated that there are half a million miles of hedgerows in England alone; this might sound like a lot, but that mileage is diminishing, despite protective measures by conservationists and the government. It's easier to mark a boundary with fence posts and wires than to undertake the laborious task of maintaining the hedgerows. Laying hedges – in which the stem of the plant forming the hedge is cut almost all the way through at its base and then bent over and woven between wooden stakes – is a skilled art, not easy to master, time-consuming and therefore expensive.

The hedgerow is a piece of living architecture and archaeology, telling a tale of human progress and endeavour throughout many centuries. The Romans planted thorn hedges in Britain to mark property boundaries; a good choice, since thorns are fast growing and difficult to penetrate. There is also evidence of ancient Anglo-Saxon hedgerows and, in some places, the boundaries they marked are still in existence today. Hedgerow planting really took off, however, as the population expanded from the 15th century onwards and agriculture became an important way of supplying commercialised food. Then, in the 18th and 19th centuries, the Enclosure Movement saw further planting of hedgerows as wealthy landowners created boundaries on common land that had previously been available to everyone. This land was then used for intense crop cultivation and for breeding sheep.

It's ironic that the movement that gave rise to most of the hedgerows we enjoy meandering along today, as an escape from the city, was at the time reviled by a rural population who no longer had the right to use the land freely. The result was that these people had to flock to the towns to find alternative means of feeding themselves and their families, an experience poignantly described in the poems of John Clare.

There are several ways of guessing the age of a hedgerow. The first is by the depth of the hedgerow; those with steep banks are a good indicator of great age. The presence of certain plants gives us a clue, too. The life history of many plants can stretch back thousands of years. Bluebells and primroses are old woodland plants and if you see them in a hedgerow, it's likely that the hedge was once part of ancient woodland felled during the Enclosure Movement and that the forebears of those bluebells picked by children today are the antique plants of many generations before.

Otherwise, there's a useful rule of thumb, published by botanist Dr Max Hooper in his 1974 book, *Hedges*: Take a 30-yard length of hedgerow and then count the number of woody species in it. The number of species equals the age of the hedge in centuries.

However, this formula tends to waver after the thousand-year marker is reached, but luckily we have some documentary history of known ancient hedges that we can use as a benchmark. And when mature trees in hedgerows fall, either through wind damage or as part of the maintenance of the hedgerow, then we have an opportunity to deduce the age of the tree (and thus the hedge) by counting its growth rings.

Next time you see a hedgerow, take time to count just how many different plants it's made up of.

Hedgerows are well known for the diversity and variety of plant species that form them, and also for the animals, birds and insects that are protected and nourished by them. Seeds blowing in the wind are stopped in their tracks by hedgerows and take root there, the young shoots thriving in a protected environment. Hedges prevent soil erosion, too, so the young plant has a dense bed of highly nutritious soil in which to flourish. We know that at least 65 species of birds live in our hedgerows and that they are crucial to the survival of moths, bats and dormice.

Hedgerows can provide a great fund of fresh, wild food for us, too, and this is partly what this book is about.

It's not the having, it's the getting
I say that is 'partly' what this book is about. It's true that there's a great deal of fun to be had in gathering

wild plants and greenery. But I'm reluctant to call this a 'foraging' book, since it's not just about identifying and collecting plants for cooking and eating, but also about communicating with, and understanding, each other and the plants and animals with whom we share our world. As the actress Elizabeth Taylor once said, 'It's not the having, it's the getting'. I don't know if Liz spent much time up to her ankles in bramble thickets and she seems to have been more into diamonds than dog roses, but nevertheless her quote is just right. A summer day exploring the woods and hedgerows with friends, family and children might yield some lovely baskets of fruit and vegetables all for free, but, for me, the real value is in spending time with one another in a very simple way, with a loosely-defined common purpose.

I want this book to be a trigger for memories: of a summer filled with hot sunny days; or of the time you got drenched in a sudden rain shower miles away from the warmth and dryness of your car or home, which made that hard-earned blackberry pie even more enjoyable. I want our children to look closely at a ladybird on a leaf, to examine and then taste the tender whiteness of a new shoot of grass, to be thrilled by the magic of an iridescent dragonfly swooping ahead on the riverbank or the sudden scurrying of a newt, or to relish the squelchy feel and juicy, rotten smell of wet mud as they fall into it.

It seems to me that the more homogenised and 'samey' our world becomes, the more our supermarkets are stocked with exotic foodstuffs from all over the world, the cleaner and more sanitised and the more 'packaged' everything becomes, the more we hanker after a return to simpler times. There is still a world where the potatoes are muddy and you might find a slug or two in your lettuce. There's a rich and delicious world out there but there's also a sort of fear – beyond the fashionable trend for foraging – about the safety of food that you, yourself, have found in the wild. Maybe we're too accustomed to our food being packaged, clean, neat and tidy? With this book I want to reintroduce the wild and natural hedgerow ingredients that our grandmothers used on a regular basis – but give them a contemporary twist.

Now, the sharper-eyed amongst you will notice that there are a couple of plants included in this book that Cynddylan's tractor might never come anywhere near during the normal course of his duties. These are samphire and laver, two of our most abundant sea plants. Although, strictly speaking, seaweeds do not belong in hedgerows, I wanted to include them here to show the versatility of British wild plants, so I hope you will forgive a little indulgence.

All the plants selected for inclusion in this book – including those seaweeds – should be familiar to most of us, common, accessible and, at times, even taken for granted. I'd like you to look again at these very obvious plants and see them in a new light. There's nothing here that will poison you or that is not easy to identify, both by adults and children. There's one more thing to add …

How to gather

Even though the plants in this book are commonplace, it's important that we preserve and respect them. There's a responsibility that comes with gathering wild plants. First, never pull up a plant by its roots – always cut above the root with a knife. Second, leave plenty for the birds. Third, never pick plants if they are scarce in a certain area. Last, never take too many plants from one particular patch. It's as well to remember that when our parents were children, primroses were so common, they were picked in huge loads and shipped to London at a premium. This doesn't happen any more, so just be aware of what you're doing and where you are.

Oh, and there's one other thing …

When you harvest any wild foods, inevitably you will harvest a number of insects and small creatures with them. It seems to be a reflex action to squash these creatures. I can't understand this, since apart from anything else they got to the plants before we did, so treat them with the respect they deserve!

Above all, have fun!

Treat the earth well.
It was not given to you by your parents;
It was loaned to you by your children.
We do not inherit the Earth from our Ancestors;
We borrow it from our Children.
AMERICAN INDIAN PROVERB

Specialist Equipment

You'll be pleased to hear that the majority of recipes in this book don't need anything that you're not likely to have in the kitchen already. However, there are some pieces of equipment that you might find handy, especially if you're going to have a go at making wines and jellies. Some of these are the sort of thing you might find at car boot sales, otherwise any decent hardware or cook shop should have them.

These include, for wine:

Demijohns (if you pick them up second hand make sure that they are washed and sterilised scrupulously before use)

Airlocks and bungs
A funnel
A good, large sieve
A straining bag

And for fruit jellies:
A large, heavy-bottomed pan (if you
  can get your hands on the 'proper'
  wide pan that's a bit like a shallow
  bucket, so much the better)
A jam thermometer
Jam jars and waxed discs
A large-sized decent sieve and a jelly
  straining bag, as above

You might also be lucky enough to come by a straining rack; this is a tripod-shaped stand from which you can suspend your jelly bag, full of fruit, over the bowl, to let the juice drip through overnight.

Also, scraps of muslin are useful for tying up herbs/flowers etc. to be cooked.

If you're sterilising pre-used jam jars, make sure there's no debris left on the insides of the lids. Then wash in hot soapy water, rinse thoroughly and leave to dry, upside down, on a baking tray lined with paper in a 170°C/gas mark 7 oven for 10 minutes. And do remember that you shouldn't put cold ingredients into hot jars, or hot ingredients into cold jars.

(Wild) Angelica

Angelica Sylvestris

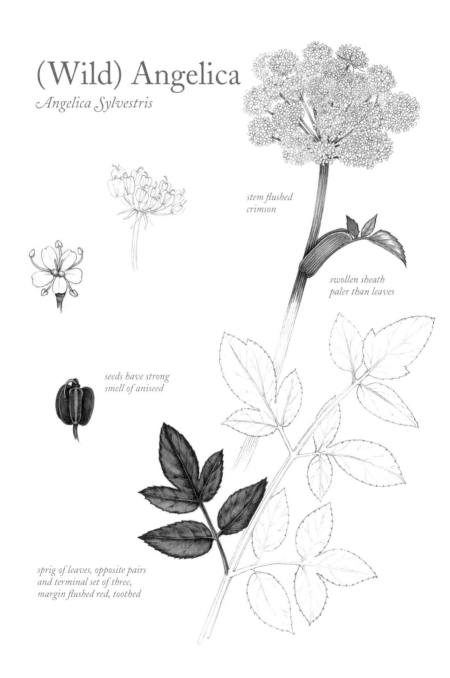

stem flushed
crimson

swollen sheath
paler than leaves

seeds have strong
smell of aniseed

sprig of leaves, opposite pairs
and terminal set of three,
margin flushed red, toothed

This tall, elegant perennial can grow to over 2 metres high. Its leaves are shaped a bit like those of the rose, with wide 'teeth'. The leaves are evenly spaced on slender stems that sprout from the stiff, upright main stem that is blotched with purple, like a birthmark. Both the leaves and the flowers – large frothing umbels – sprout from sexy, tightly folded sheaths. These flowers form an explosive star shape, the tiny blossoms tinted white and pink. The plant flowers from July to October.

Angelica likes to inhabit moist waste ground; since its seeds can float, the plant can often be found drifting alongside the banks of rivers or canals.

That beautiful name, *Angelica*, is the subject of several legends. One of these suggests that the qualities of the plant were revealed by an angel, hence the name; the plant, said the angel, could cure the plague.

Another explanation for the name is that the plant is usually in flower at the time of the feast day of that most mighty of archangels, Michael, on 8 May.

The cultivated angelica, known by the exotic and glamorous Latin name of *Angelica archangelica*, can be differentiated from the wild *sylvestris* variety shown here since it has larger flowers, each umbel measuring approximately 25cm across. However,

this plant often escapes from the boundaries of its cultivation and is sweeter than the wild kind. (If you find it, keep a note of its location.)

Culinary uses

Young angelica shoots taste a bit like celery and can be used in salads and soups. Tougher, more mature angelica stalks can be boiled and used as a vegetable – if they're very tough you'll need to boil them twice. Otherwise, the candied stalks of the plant are often used as a sweet or added to cakes and desserts.

Angelica seeds, which are abundant, were once used to flavour vermouth, gin, and most noticeably, chartreuse. If you want to harvest them, gather the seed heads on a dry day and put them on a white cloth (which will make them show up better) in a warm, dry place. The seeds can be shaken out cleanly and easily and stored in an airtight jar.

Medicinal uses

Every single part of the plant was once believed to be effective in healing something or other and, what's more, it was also used in magical spells to banish unwelcome ghosts and spirits.

And what about more practical modern-day applications? For home use, the easiest thing to make is an infusion of the leaves, with honey added to taste, which will ease tummy pains and help prevent flatulence. Bellyache weed, one of the old names of the plant, reflects this use.

Did you know?
Serving nibbles of the candied roots with an alcoholic drink might have a homeopathic effect, since angelica is believed to relieve a craving for alcohol!

Candied Angelica Stalks

This whole procedure takes time but it's worth it; candied angelica is delicious and can be stored in airtight jars. You can chop it into finer pieces and add to cakes and puddings or whizz into smoothies as a secret ingredient.

You will need a large, flameproof cooking pot or saucepan with a lid and a large, sealable jar.

No quantities are given for the ingredients here, and the reason will be evident when you read the recipe.

Angelica stalks
Granulated white sugar
Fine caster sugar

For this recipe, harvest only the freshest young stalks; the older ones tend to get tough. Chop the stalks into lengths of between 5 and 10cm. Put the stalks into a pan with just enough water to cover them, put a lid on the pan and boil for about 10 minutes, or until tender (the younger the stems, the quicker the cooking time). Drain the stalks, reserving the cooking water, and allow to cool before peeling the stems.

Boil the peeled stems once more in the reserved cooking water, until they are a green colour. Drain, then blot dry on kitchen paper.

Weigh the stems, and add the same quantity of granulated sugar to them (e.g. 450g sugar to 450g stems).

Let the stems and sugar stand for a couple of days in a covered bowl, then put the stems and sugar into the pan once more, again just covering with water. Put the lid on the pan, and bring to the boil, simmering until the stems are translucent and bright green. Drain, then roll the stalks in fine caster sugar (grind the sugar in a food processor if necessary) and place on a lined baking tray in a very cool oven – put the oven on its lowest setting. The slow cooking of the stalks can take up to three hours, but keep checking so that they don't get too hard. Test from time to time to make sure they retain a 'chewy' texture.

Remove from the oven and allow to cool completely. Store in a sterilised, airtight jar.

Angelica Seed Vinaigrette

The flavour of angelica seeds gives an intriguing piquancy to salad dressings.

1 tbsp olive oil
Zest and juice of ½ orange or lemon
Splash of white wine vinegar
1 tsp wild angelica seeds

Simply whisk all the dressing ingredients together and add to a salad of watercress or any other fresh leaves you have to hand. It's best to use straight away, but can be stored in an airtight jar in a cool place for a couple of days – not in a fridge, otherwise the olive oil will coagulate.

Ash
Fraxinus Excelsior

*leaves flat, distinct
paralell veins*

*just emerged leaves
very pale*

*ash keys
in big bundles,
becoming brown with
age, and the twist
increases as the key
dries out*

This is a truly magnificent tree, which can attain both a great height (up to 46 metres) and a great age. It's with good reason that it's called the Queen of the Forest, the counterpart to the oak as the King. The bark of the ash is initially pale and smooth, becoming more fissured with age. The leaves of the tree are distinctive, a little like rose leaves, with 7–13 symmetrically-spaced leaflets along either side of the stem. The flowers, which appear before the leaves unfurl, are both male and female on the same tree. The blossoms are tiny bundles of leafless purplish dots; the female flower heads are longer than those of the male.

Culinary uses

Those Mediterraneans … they've got all the good stuff, right? Spices. Olives. Capers. All the lovely things that lend an exotic touch to everyday food and tingle your taste buds with delight. It's a real shame, then, that not many people know about ash keys, which certainly rival the more exotic foods in unusual tastiness. They're simple to prepare and, served as hors d'oeuvres or added to pasta and salads, will keep your dinner guests guessing.

The ash tree is of the genus *Oleaceae*, the same as the olive tree. So it's not really surprising that its seeds, also known as 'samara', have an intensity that rivals that of olives and capers.

The 'keys' we're talking about stay on the tree right through the autumn and winter, and only fall to the ground the following spring. But the time to harvest them is when they're young and fresh. There are several recipes for pickled ash keys but the one here has an unusual twist. It calls for West African curry powder but you could use Indian curry powder.

Medicinal uses

Opinions change quite rapidly as to what is actually considered 'medicinal'. A couple of hundred years ago, the ash was thought of as a healing tree, its powers almost magical. But some of those old customs and traditions would qualify as mere superstition today. Older ash trees often develop large vertical holes in their trunks, big enough to climb into and out of, and children would be passed, naked, through this gap. This was meant to fix a number of ailments, but was regarded as a cure for weak limbs and bones in particular. This might have been in the hope that the child's bones would take on the strength of the tree itself. I've seen children – and adults – climbing through the vertical hole in an old ash tree that lives close to me but I doubt they realise that they're instinctively copying a very old ritual.

Did you know?

In Norse mythology the great World Tree, or Tree of Life, is defined as an almighty evergreen ash, Yggdrasil, whose roots extend to the depths of the Earth (Hel), then come up through the middle of the Earth (Midgard) and whose branches disappear into the highest heavens (Valhalla). The tree is watered by the twin concepts of faith and wisdom.

Ash wood is not only strong but very elastic, which means it's an ideal timber for all sorts of uses and is particularly favoured for making the bodies of guitars and drums.

Spicy Pickled Ash Keys

As well as a large, sealable jar you will need a fine sieve lined with muslin or cheesecloth.

MAKES 1 X 500ML JAR

400g young green ash keys
2 tbsp salt
6 black peppercorns
4 dried piri-piri chillies
12 mustard seeds
4 garlic cloves, peeled

SPICED VINEGAR
450ml white malt vinegar
1 tbsp paprika
2 tsp cayenne pepper
1 tbsp West African curry powder
Salt, to taste

First, make the spiced vinegar. Mix all the ingredients together in a saucepan. Bring to the boil and immediately take off the heat, then leave to cool. Set aside.

Detach the stems from the ash keys and clean the keys thoroughly. Pop the keys in a large pan and cover with water. Bring to the boil, then cover and simmer for 20 minutes. Drain the ash keys, put them back in the pan and cover with water. Bring to the boil again, cover and simmer for 20 minutes, then drain. Repeat this process twice more. It might sound like a bit of a bore, but if you don't carry out this process you'll probably find that the keys are unacceptably bitter.

Once you've drained the ash keys for the last time, add them to the pan once more, cover with water and add the salt. Bring the mixture to the boil and continue boiling briskly for 15 minutes. Reduce to a simmer, cover and cook gently for 1 hour. After this time the keys will have softened and the bitterness will have disappeared.

Drain the ash keys and return to the pan.

Strain the spiced vinegar through a fine sieve lined with muslin or cheesecloth, then add to the pan with the ash keys, and add the peppercorns, chillies, mustard seeds and garlic. Bring the mixture to the boil, reduce to a simmer and cook for 10 minutes.

Take off the heat, allow to cool, then transfer the ash keys to a sterilised jar. Cover with the spiced vinegar, seal, and set aside to mature in a cool, dark cupboard for at least 6 weeks.

Beech
Fagus Sylvatica

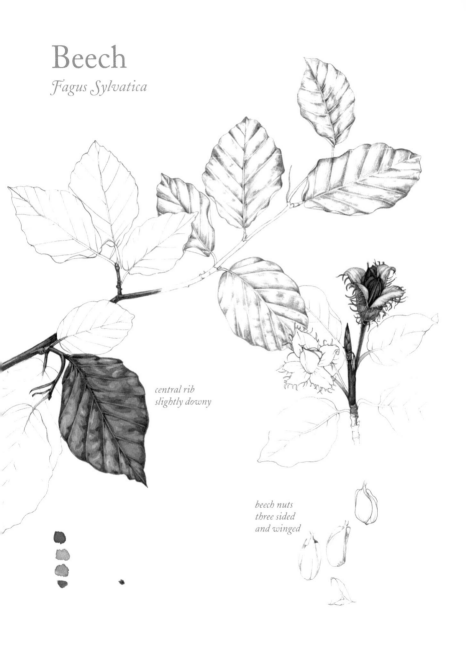

*central rib
slightly downy*

*beech nuts
three sided
and winged*

The beech tree is arguably one of our most beautiful and graceful trees, although it's so common that we sometimes don't notice its fragile, elegant loveliness. In the spring, beech leaves are a delicate lime green tone, soft and smooth to the touch and almost translucent. By summer, though, the leaves have turned a darker green, and in autumn they dry to a delightful coppery colour, the same tone as truly auburn hair. The fruit of the beech tree, beechnuts, also known as mast, are borne inside a hard cup-like shell with 4 hairy, prickly sides that change from green to brown as they ripen. In autumn, the husks burst open and fall to the ground, releasing the nuts.

Beech trees are tall (30 metres or more) with smooth bark, almost a metallic grey in colour. The leaves curve to a sharp point and have slightly serrated edges. If you look at the leaves from their sides, along the edge, you'll see that they are slightly wibbly. It's easy to mistake a beech for a hornbeam; beech leaves are shinier on top. The tree is deciduous.

Culinary uses

Both the leaves of the beech and its nuts are edible.

Beech leaves: The tender young leaves make a great addition to salads; however, there's another way of using them, which is way more interesting, and that's to make an exotic liqueur, as in the recipe below.

Beechnuts: The furry husk of the beechnut – also known as 'mast' – contains 3 small triangular-shaped nuts. When the husk is empty it has a pretty flower-like appearance and is handy for use in flower arrangements if you're into that sort of thing. The nuts themselves are delicious, but it makes for painstaking and very fiddly work to extract them from their hard, brown protective skins and for that reason they're not really commercially available.

Roasting the nuts makes them easier to peel, after which they should be rubbed to make sure any small hairs have been thoroughly removed.

They can then be eaten raw, or if you prefer the taste of toasted nuts, toast them in a pan over a high heat, shaking constantly so that they don't scorch. Don't worry about adding oil. The toasting process, if you've never done it before, tends to be faster than you might think, so be aware!

You can add the nuts, whole, to crumbles or anything else that needs a crunchy topping. They can also be used to replace any other nut in a recipe, but because of the trouble you have to take to get at these tiny delicacies, I thought it would be nice to make something a bit special with them, a turron-style nougat. Turron is stickily delicious and a special Spanish Christmas delicacy.

Medicinal uses

An old remedy to counteract rheumatism was to drink beech tea, with added lard.

Did you know?

It's an old superstition that beech bark is deadly poisonous to snakes.

Beech Leaf Noyau

It's lovely to hold back a few bottles of this exotic-tasting aperitif for Christmas time, and then try to get your friends to guess what it is. You can drink it neat, or over ice, or with tonic water or soda.

FOR EVERY 70CL BOTTLE OF GIN YOU WILL NEED:
400g fresh young beech leaves, with stems stripped off
225g granulated white sugar or caster sugar
300ml water
200ml brandy (or cognac, or similar)

Put the gin and the leaves into a large, sterilised glass jar, seal and leave to sit for 3 weeks. After 3 weeks, strain the gin from the leaves. Boil the sugar in the water. Allow to cool, then mix together with the strained gin and the brandy. You can then decant the mixture into attractive bottles, and store.

Beechnut Turron

I can't stand the taste of the blended, heated type of honey and, in this recipe, I like to use a single varietal Spanish honey since the nuts deserve a special partner.

In total you will need 1.5kg assorted nuts, half of which should be beechnuts and the other half whatever nuts you like, such as almonds or filberts. Half the entire quantity of nuts, whatever type they are, should be ground, so you could maybe buy ground almonds to make life easier.

MAKES APPROX. 2KG

1kg honey
500g white sugar
2 egg whites
750g beechnuts
750g assorted ground nuts (e.g. ground almonds or filberts)
Zest of 1 lemon
Rice paper or wafer paper (optional)

Heat the honey in a heavy-bottomed pan over a low heat, taking care that it doesn't burn. Add the sugar to the pan and stir with a wooden spoon until the sugar has dissolved. Remove from the heat and allow to cool a little.

Whisk the egg whites until they form stiff peaks (the usual test is to hold the bowl upside down over your head – make sure you have an audience just in case there's anything to laugh at) and add to the honey and sugar mixture. Stir for 8–12 minutes before putting the pan over a low heat again. Stir constantly until the mixture caramelises.

Add the nuts, together with the lemon zest. Mix everything together thoroughly, then fold into a shallow baking tin lined with greaseproof paper. The turron will need to be about 2.5cm deep. Leave to cool for 2 hours. If you're using wafer papers, line the bottom of the tin with them first. You can also lay papers over the top, too, if you like.

Once the turron has completely cooled, it is ready to eat and you can score it into squares, or bars, or whatever you like. If you're the optimistic sort of person who thinks you're going to store your beechnut turron, it will keep well in an airtight container.

Birch

Betula Pendula

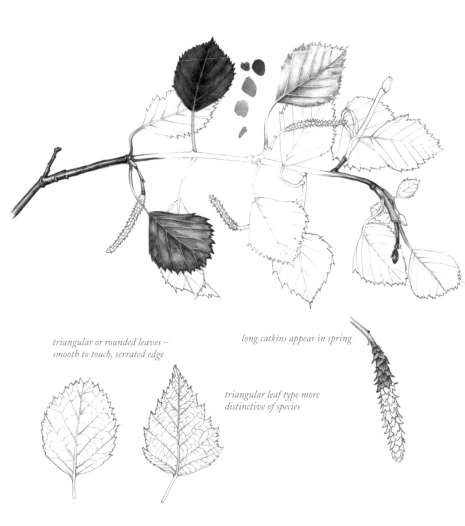

*triangular or rounded leaves –
smooth to touch, serrated edge*

long catkins appear in spring

*triangular leaf type more
distinctive of species*

The birch we're talking about here is the silver kind and, although there are other types, all serve the same purpose in terms of edibility.

A very elegant deciduous tree, the silver birch is so-named for its silvery white bark, which is broken up by a design of horizontal black lines. The leaves are diamond-shaped, smooth to the touch, with jagged edges. The birch tree has long catkins that appear in the spring. Birches don't grow to be massively tall – 25 metres at most. Birches grow quickly and easily and, despite their ethereal beauty, their rapid growth sometimes makes them unpopular with those who consider them to be 'weeds'.

Culinary uses

Of all the ingredients obtained from trees, birch sap is one of the most intriguing. To our ancestors, whose access to sweet stuff was far more limited than our own, the deliciously fragrant and delicately flavoured sap of the birch tree would have been very welcome after a long winter when whatever food was left in the stores would have become stale and boring.

Everyone should try tapping a birch tree at some point, but there are a couple of key points to bear in mind.

First, you should never try to tap very young trees since this can damage them. To determine whether the tree you've pinpointed is old enough, simply measure its diameter – if it is more than 30cm, the tree is mature enough to tap. Never tap the same tree more than once in a 3-year period. Another very important point is to close up the hole that you make in the tree, otherwise the tree can, quite literally, bleed to death. The best time to harvest birch sap is early spring, as soon as the fresh young leaves are making a decent show – usually in early March. You'll need a drill with a sterilised 10mm drill bit, a 5-metre length of virgin tubing, a large sterilised container to collect the sap in and a sterile cork to plug up the hole you've made.

Drill the hole in the trunk about 1 metre from the ground, angling the drill slightly upwards. The hole should be approximately 50mm in diameter. Push the tubing into the hole, with the other end placed in your sterilised container (the sort of glass demijohn used for making wine would do the job nicely). Stuff some clean cotton (an old clean tea towel cut into strips will do) into any spaces between the top of the container and the tube, and also make sure that there are as few gaps as possible between the tube and the tree otherwise you'll lose a lot of sap.

Then you simply need to wait. Don't be greedy; to be on the safe side, only take 4.5 litres of sap per tree.

Once it's collected, the sap won't last very long, so either harvest it as you need it or pop it in the freezer.

You can use the sap to make birch syrup – but it's not advisable because of the huge volumes of sap you'd need to collect, since approximately 91 litres of sap reduces down to 4.5 litres of syrup. A tried and tested, timeless use of birch sap is for making birch sap wine.

Medicinal uses

The dried young leaves of the birch are slightly antiseptic and also diuretic, so are used to treat urinary tract infections. If you want to try this, just add a handful of leaves, and some sugar to taste, to a teapot, leave to infuse and then strain into a mug.

Did you know?

Birch bark, which peels away easily from the tree, was used by ancient Britons and also Native Americans to make canoes. This same bark has also been used for thousands of years as a kind of 'paper'; rolls of it have been found in Mesolithic excavations. At that time, writing was a magical act, the skill of the priest or shaman. The bark has fungicidal properties, which explains its longevity.

In less enlightened days, to be 'birched' was the dread of any schoolchild. The flexible, long, young branches were used as a whip to punish the young transgressor.

If you're making an omelette, and you have a birch tree handy, try using a bundle of the twigs as a very effective and unusual whisk. Strip off the bark from the twigs – this is easy – and tie them together with a piece of dampened jute string. As the string dries it will tighten. *Voilà*, a free kitchen implement!

Birch Sap Wine

To make this wine, you will need a large food-grade bucket and 2 demijohns with airlocks. Much of this equipment can be a lucky find at a car boot sale or charity shop. You will also need some sterilised bottles with corks.

MAKES APPROX. 3 X 70CL BOTTLES

4 litres birch sap
1kg granulated white sugar
200g raisins
Juice of 2 lemons
1 7g sachet wine yeast, made according to the manufacturer's instructions

Use the sap as soon as you have it – it goes off very quickly. Put it into a pan and bring to the boil. Add the sugar and simmer for about 10 minutes. Put the liquid into the food-grade bucket and add the raisins and lemon juice.

Make up the yeast according to the manufacturer's instructions and, once the sap mixture has cooled to blood temperature, add the yeast to the bucket. Leave the liquid to ferment for about 3 days.

Once it has fermented, strain the liquid into one of the demijohns and seal with an airlock. Leave the demijohn in a warm, dark place, such as an airing cupboard, until the fermentation is finished (you'll be able to tell when this happens since no more bubbles will be rising up into the airlock, but it can take up to 8 weeks). Then take the second demijohn and carefully filter off the liquid, leaving behind all the sediment that should have settled in the bottom.

You might need to carry out this procedure again before bottling. The wine is generally at its best after a year. Leave the bottles on their sides so that the wine can further mature in the bottle, but make sure that the corks are tied on tightly.

Bird Cherry
Prunus Padus

light green leaves
with regular teeth

purple-black fruit,
one stone per fruit

stem red

This plant appears either as a shrub or as a small tree that grows to about 15 metres in height. In contrast to other types of cherry, the blossoms of the bird cherry hang in drooping, dangling lengths, giving rise to its folk name of 'wild lilac' in some parts of Yorkshire. The fruit that follows this blossom appears from July onwards.

The tree thrives in dampish areas such as the edges of streams or rivers or wide ditches and can also be found in wet wooded areas.

Culinary uses

The fruits of the bird cherry – sometimes called 'hags' – are much more bitter than regular cherries, but the birds don't seem to mind this and gobble up the fruit as soon as it appears, hence the name. The fruit is more commonly used in parts of Central Europe, but since there are plenty of bird cherries in the UK it seemed a good idea to include some ideas here. Bear in mind that acidic taste and compensate for it with plenty of sugar.

The kernels of the bird cherry 'stone' should not be eaten raw – although it's unlikely you'd consider doing so – since, in its uncooked state, it contains some toxic compounds. However, cooking the stones (as in the recipe for bird cherry flour) makes them safe to eat.

Medicinal uses

If you were living in the Middle Ages then it's likely that you might have tied up a piece of bird cherry bark above the doorway to your house in the hope that this would ward off the plague. Sadly, we have no evidence to tell us whether this was effective or not. On the whole, given the range and ferocity of the plague, I would guess not.

Did you know?

The wild cherry tree, *Prunus avium*, is often mistaken, in name at least, for the bird cherry, because of the 'avium' part of the name, which means 'bird'.

Bird Cherry Brandy

The astringency of the bird cherry 'hags' lends a delicious and unusual flavour to this brandy: the perfect warming tipple on a cold winter evening. You can also add a dash of this brandy to mulled wine.

MAKES APPROX. 1.5 LITRES

500g cherries, stoned
1 litre brandy
225g caster sugar
200g blanched almonds

You'll need a lidded container large enough to take all the ingredients, one that's made of a material that won't contaminate the flavours – glass or pottery is ideal.

This complex and demanding recipe stipulates that you plonk all the ingredients in the jar and then give it a 'hello there' jiggle every time you pass it by. After 2–3 months, press the cherries and almonds through a sieve and bottle the resulting delicious cherry brandy. Don't be tempted to leave the cherries any longer than 3 months otherwise they will start to discolour, and the taste will be too strong. Add the left-over boozy cherries to a pie or crumble along with plenty of sugar.

Bird Cherry Flour

This is a very unusual recipe for the UK but not at all uncommon in Russia. If you are lucky to have a bird cherry tree nearby you will know that they yield fruit in abundance, so it's worth the effort to give this a go.

You will need a large stone pestle and mortar – and plenty of elbow grease!

MAKES APPROX. ½ KG

1kg bird cherries, left whole

Wash the cherries. Smash up the cherries, stones and all, in batches, in the old-fashioned way in the pestle and mortar (hence the need for elbow grease). Make sure the stones are in pieces. Then take 2 large baking trays, line with baking parchment, spread all parts of the cherries on them and put into a cool (110°C/ gas mark ¼) oven. Leave the cherries to dry out – this will take several hours. Turn the fruits from time to time to make sure that they cook evenly and don't burn. When you consider they are done (i.e. all parts of the cherries are dry and brittle), leave them in the oven but turn the heat off and allow them to become completely dry.

Leave overnight before grinding finely in a coffee or spice grinder. Store in an airtight container.

You can use 50/50 'normal' flour and bird cherry flour in any number of recipes – it lends a delicious and chewy texture to your baking.

Blackberry
Rubus Fruticosus

fruit starts bright green,
flushes crimson and finally
becomes deep purple black

pink petals, some still
apparent with friut,
although withered

leaves in sets of three,
occasionally five

If you've been for a country walk through woods or across scrubby heathland just about anywhere in the UK, it's likely that you will have snagged your clothes on the long, whippy, thorny creepers of the blackberry or bramble bushes that arc persistently through the undergrowth. The pretty pink-white blossoms are borne on these long stems between May and September, followed by the shiny purple-black fruits that can stain your clothes, your hands – and your teeth! Some people use the old country name 'bramble' for the blackberry. This name comes from the Germanic word 'brom', which means 'thorny shrub'. Blackberries are one of the most prolific fruits in the British Isles, instantly recognisable to everyone, probably, and popular both because of their abundance and their deliciousness. There's a decently long harvesting time, too, from July to October, depending on where you are in the British Isles, although the first frost does tend to make the berries tasteless and inedible. For the best flavour, pick the berries towards the end of a hot summer afternoon, when the sunshine will have given the fruits the juiciest, sweetest taste.

Culinary uses
There are dozens of ways to use blackberries. You can make blackberry jam, tarts, pies and even wine. They're delicious eaten alone, or stewed and dolloped on top of yoghurt, or layered into bread and butter pudding as a surprise ingredient. I like to add frozen blackberries to my winter porridge – they soften immediately, leaving lovely purple streaks and a welcome hit of vitamin C.

Medicinal uses
An infusion of the leaves of the blackberry eases a sore throat; either drink it (it's very nice and makes a good substitute for 'normal' tea in any case) or gargle with it.

Did you know?
It was considered very bad luck to eat or collect blackberries after 10 October because the Devil apparently spat on the bushes on the previous evening!

Tip
Before using blackberries, leave them on a large sheet of baking paper for half an hour so that any insects can escape. Then put the berries in a colander. Sprinkle the surface with salt. Put the colander in a large bowl of water, just to the level of the berries. This should bring any remaining insects to the surface.

Chilli Blackberry Syrup

This recipe was kindly sent to me by Andi Oliver, chef, singer, TV presenter and all-round Renaissance woman. If you have plenty of blackberries to hand, you really must try it. It's quick to make and you'll be left with enough sweet and spicy, chilli-infused blackberry syrup to keep your taste buds tingling right into the winter if you're smart enough to freeze some. You can use it to spritz up sparkling water, otherwise it's also great swirled into yoghurt, oatmeal and crème fraîche; you can slather onto buttered toast, drizzle over goat's cheese, or use as a flashy, unexpected topping on pancakes, crêpes, or waffles.

MAKES APPROX. 500ML

4 dried guajillo chilli peppers or 25g any
*   other chilli pepper you care to use (hot*
*   is good since this is, after all, a spicy*
*   syrup)*
Good glug of pomegranate molasses
170g muscovado sugar or dark brown
*   sugar*
200g organic white sugar
350ml water
Juice of 1 lemon
100g blackberries, washed and with
*   stalks removed*

Trim the stalks from the dried chillies. Tear into pieces and drop (along with the seeds) into a medium saucepan. Stir in the molasses, sugars, water and lemon juice, and bring the mixture to a boil over a medium heat. Boil, stirring regularly, until the mixture has reduced to 475ml. This will take about 20–30 minutes.

In the meantime, purée the blackberries in a blender. Then force the berries through a fine-mesh strainer and discard any seeds. Set the berry purée aside.

Once the chilli mixture has reduced, remove it from the heat, and (carefully) purée it with a hand blender until smooth. Strain through a sieve into a heatproof bowl. Press on the remaining solids in the sieve to squeeze out any syrup, then discard the remaining mush.

Whisk the berries into the chilli syrup and set aside to cool. When cool, spoon into a large, sterilised jar, or several smaller jars, seal and refrigerate.

Blackberry Cordial

MAKES 600ML

*1.8–2.25kg blackberries, washed and
  with stalks removed
225g granulated white sugar
2–4 cloves
1 stick cinnamon
Rind of ½ lemon*

Put the blackberries into a saucepan
without any water. Heat them slowly
until the juice is oozing, being careful not to boil. Strain through a jelly
bag or muslin to remove the pips.

Return the juice to the pan, add the
remaining ingredients and boil for 30
minutes. Strain again, then pour into
sterilised bottles, and seal.

To use, add water as you would
with any other cordial. Alternatively,
try it with vodka or gin, or add to
Champagne or Prosecco as a delicious alternative to cassis.

Blackberry Sorbet

SERVES 4 ADULTS, OR 2 CHILDREN!

*100g granulated white sugar
150ml water
450g blackberries, washed and with
  stalks removed
White of 1 small egg*

Boil the sugar and water for about
5 minutes to make a syrup. While it's
cooling, sieve the blackberries to get
rid of the pips, then add the blackberries to the syrup. Beat the egg
white to soft-peak stage and fold it
carefully into the mixture with a
metal spoon. Freeze for about half an
hour to a slush, stir, and freeze again.
In a normal freezer it will take approx. 2-3 hours to make, but if you
have a 'fast freeze' button on your
freezer, the whole process will take
about half the time.

Blackberry and Crab Apple Jam

No self-respecting book about hedgerows would be complete without a recipe for gorgeous blackberry jam. The fruit is so abundant that you might want to consider making extra jars for your friends. I have also found that the cheeky addition of a glug of whisky, to replace some of the water, lends a certain unexpected pleasure.

You will need some warmed, sterilised jars and wax discs.

MAKES APPROX. 5KG JAM

2kg blackberries, washed and with
  stalks removed
1kg crab apples (or sour baking apples),
  cored and chopped
300ml water
3kg granulated white sugar
Good glug of whisky (optional; if using,
  reduce the amount of water by the
  amount of whisky used)

Simmer the blackberries in a large pan with half the water, until they're soft. Do the same with the apples in a separate pan, using the other half of the water. Mix the 2 fruits together in one pan, remove the pan from the heat, allow to cool slightly and stir in the sugar. Stir until it has dissolved, bring to the boil and allow to boil vigorously until setting point is reached; you can determine this by putting a small blob of the jam on a chilled saucer. If the jam's ready, it will wrinkle if you push it with your finger. Pour into warmed, sterilised jars, cover with wax discs and, when completely cold, put the lids on.

If you want to make a smooth jam, then you can sieve the cooked blackberries before mixing them with the apples. You'll need slightly less sugar if you do this – just 2.25kg. But it's such a messy process, I'd suggest you live with the pips!

Blackberry Muffins

MAKES 12

250g self-raising flour
1 tsp baking powder
50g butter, at room temperature
50g caster sugar
175g blackberries, washed and with
* stalks removed*
Juice of 1 lemon
Grated zest of 1 unwaxed orange or
* lemon*
2 eggs
150ml milk

Preheat oven to 200°C/gas mark 6;
muffins like a hot oven from the
start. Pop your paper muffin cases
into a 12-hole muffin tray.

Sift together the flour and baking
powder; this makes the muffins
airier. Then rub in the butter. Add
the sugar, stir in the blackberries,
lemon juice and orange or lemon
zest. Mix gently with your hands.

In a separate bowl, beat the eggs
and milk together, and then add to
the flour mix. Stir together quickly
with a fork – you're not aiming for a
particularly smooth batter but you
do need to work fast. Dollop the
mixture into the paper cases and
bake for about 20–25 minutes. Leave
to cool slightly before serving with a
nice cup of tea.

Blackberry Leaf Tea

Finally, it's not all about the berries!
Blackberry leaves make a lovely tea,
which not only tastes good but also
eases a sore throat. It's best made
with fresh young leaves gathered in
the springtime. Simply throw a
(washed) handful into a teapot,
steep, and add sugar to taste.

Borage
Borago Officinalis

*stem at top of flowering
area flushed purple*

*fruiting
capsule*

beautiful blue sepals

*whole plant bristly,
stout white hairs*

Borage is sometimes called 'starflower' and this name reflects the borage flower's shape and its pointy petals. Borage is quite a tall plant, growing to a height of approximately 60cm, and the leaves and stems are distinctly hairy, although not prickly. The leaves are oval, with crinkly edges and a 'quilted' appearance. The borage flower is a most heavenly blue colour.

Culinary uses

All parts of the borage are edible, although for some reason people seem to have forgotten about its usefulness of late. If you read any historical accounts of the borage, one thing strikes you: the plant's association with happiness and its ability to chase away the blues ... or 'melancholia' as our great-grandparents would have called it.

Borage leaves – and stems – taste a bit like cucumber, soothing and refreshing. You might be put off by the tiny downy hairs; don't be, they just sort of dissolve on your tongue.

You can use the leaves in salads and soups, add them to sandwiches instead of cucumber, or chop them up and add to plain live yoghurt to make a raita. Use them instead of cucumber in Pimms. If you have bundles and bundles of borage – which, after all, can spread abundantly – you can strip the leaves and use them in the same way as spinach.

Throw a handful of leaves into a teapot, infuse, and *voilà* – borage leaf tea. Add sugar, chill and then throw in a couple of mint leaves and a slice of lime for a gorgeous summer drink served over ice.

If you've a deep fat fryer or a decent wok, take some borage stems, dip them in tempura batter, and fry fast. This tasty snack is popular in some parts of Spain. There's a recipe for the batter on page 143 in the ox-eye daisy section.

Borage flowers, too, are edible. They look beautiful thrown into a salad just before serving. You can also freeze them in water in ice cube trays to make a stylish addition to summer cocktails. If you're going to do this, use bottled water since it tends to make clearer ice than tap water. Another way of showing off the flowers is to crystallise them: leaving the stems on the flowers, coat them in a little whisked egg white, then dip the flower carefully in fine white sugar and leave to dry overnight.

Medicinal uses

Any plant with *officinalis* in its Latin name is a clue that it is considered to be a medicinal plant. The dried flowers and leaves of the borage are used as a diuretic and, in years gone by, were considered effective in the treatment of inflammations of the kidney and bladder. Borage tea is used to ease rheumatism, and asthmatics might like to try it since it strengthens the lungs. It's used, too, for other respiratory infections, such as bronchitis.

Did you know?

It was once believed that the borage made you brave; the Celtic word 'borrach', meaning courage, could be the origin of the name. Warriors were given drinks containing borage flowers before they went into battle.

Claret Cup

This old-fashioned recipe comes courtesy of the inestimable Mrs Beeton. The claret cup is passed from person to person – which is a lovely way to celebrate a christening, engagement or birthday.

MAKES 10 STANDARD 5OZ WINE GLASSES

1 bottle claret
1 bottle soda water
1 liqueur glass maraschino liqueur or
  cassis
4 tbsp caster sugar
¼ tsp grated nutmeg
Approx. 225g crushed ice
Sprig of borage, to garnish

Place all the ingredients into a polished silver cup with a glass lining, regulating the amount of ice according to the state of the weather (if it's a very warm day, of course you'll need more). Garnish with a sprig of borage. Pass the cup around with a clean napkin slipped through one of the handles, so that the edge of the cup may be wiped clean after each guest has had a sip.

Broom

Cytisus Scoparius

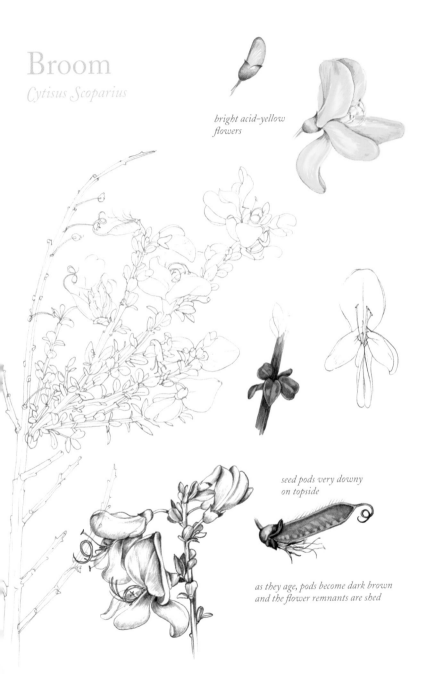

bright acid-yellow flowers

seed pods very downy on topside

as they age, pods become dark brown and the flower remnants are shed

Scottish broom is a neat shrub, upright in appearance, with long, straight stems. It can grow up to 3 metres tall. Those stems each have 5 angles, and short oval leaves. The shape of the bush is very distinctive, as is the soil type it prefers; broom generally always grows on acid soils and is therefore happy on scrubby, gritty land. The flowers, which appear in the summer, are a bright acid yellow and the seed pods that follow are black and covered in hairs.

Broom was once commonly used to make ... brooms! The long, flexible stems are perfectly suited for sweeping. However, there's an old superstition that says a besom made of flowering broom blossoms would sweep away the head of the house.

The bright, metallic yellow flowers of the broom, which flower between March and June, are very eye-catching. Broom's generic Latin name, *planta genista*, is the reason why Henry II's line were called the Plantagenets, since he wore a sprig of these bright flowers in his helmet as a means of identifying himself in battle.

Incidentally, it's important to tell the difference between the 'true' broom that we're talking about here, which has small, oval leaves clustered in bunches of threes. There's another variety, Spanish broom, which has leaves a bit like pine needles. The flowers of the Spanish broom are mildly poisonous so the plant is best avoided. Another way of identifying Spanish broom is that it flowers later in the season, from late June to the end of August.

Culinary uses

The buds of the broom flowers can be eaten as they are, and a handful thrown into a salad or stirred into cooked pasta not only looks amazing but tastes great, too – they have a perky, nutty flavour. The buds can also be pickled. Prepared in this way they taste rather like capers and this was a very popular way of using them in the 17th and 18th centuries.

Medicinal uses

Broom tips were once used as a mild diuretic. In fact, broom was gathered for this very use during the dark days of the Second World War. Drugs made from compounds in the plant are used to treat heart disease and also ailments of the liver and kidneys.

Did you know?

Although broom flowers are bright yellow, they make a green dye; hence, one of its folk names – dyer's greenweed. The bark was once one of the constituents used to tan leather.

Pickled Broom 'Capers'

This recipe is based on one by Charles Carter in *The Compleat City and Country Cook*, 1736.

MAKES APPROX. 200G

300ml cider vinegar
80g salt
200g broom flower buds, collected before
   the tops turn yellow

Heat the vinegar and salt in a pan until the salt dissolves. After rinsing them, put the whole broom buds in a large jar and top with the warm vinegar mix. Cover the jar and stir every day for a week, then transfer the broom buds to a sterilised jar, top with the vinegar mix, seal and store in the fridge for at least 2 weeks to mature before using.

They can be used in the same way as capers; stirred into pasta, thrown into salads, added to pizza after cooking or simply eaten as hors d'oeuvres.

Chamomile
Chamaemelum Nobile

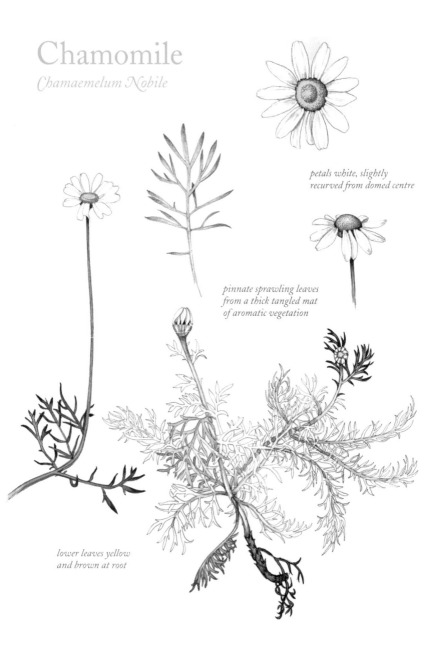

petals white, slightly
recurved from domed centre

pinnate sprawling leaves
from a thick tangled mat
of aromatic vegetation

lower leaves yellow
and brown at root

The variety shown here is also called Roman chamomile. It looks a bit like a daisy, with its yellow centre and white petals, but grows much taller in its wild state. Chamomile has wiry, bendy stems and feathery leaves and a spreading habit. The name means 'earth apple' and this describes both the habit and the gorgeous, sweet smell of the plant perfectly.

Culinary uses

There's not a huge range of things that you can do with chamomile, but so many of us buy the mass-manu-factured tea, when the plant is so abundant and the tea so easy to make from it, that I thought it worth including here.

The tea is simply an infusion of the fresh (or dried) flowers, made to taste – nothing fancier than that. Chamomile tea has a relaxing effect and aids sleep.

To dry chamomile, pick on a hot summer's day, keeping some of the long stem. Tie in bundles and hang upside down, in a warm dry place, in large paper bags, which will catch any falling leaves, petals or flowers. When dry, pick off the flowers and store in a dark jar.

Medicinal uses

As a healing herb, chamomile has wide-ranging and far-reaching uses; the Ancient Egyptians thought it so efficacious that they dedicated it to their gods. Everyone knows about its sedative qualities, but in addition chamomile is mildly antiseptic, can cure nausea, alleviate menstrual pain and helps to stimulate the appetite.

Did you know?

The chamomile flower is the 'secret' ingredient used to flavour the Spanish sherry, manzanilla, whose name means 'little apples'.

Spicy Chamomile Chai

Here's a twist on that famous tea. This chamomile chai, inspired by Indian chai, with the addition of a few spices, is a wonderfully soothing and comforting drink.

MAKES 3–4 MUGSFUL

2 tsp dried chamomile flowers or twice the amount fresh
3 tsp fresh ginger root, grated
1 tsp coriander seeds
A pinch of cinnamon
A few specks of cardamom
A soupçon of allspice
600ml (1 pint) milk
Soft brown sugar, to taste

Make a bundle by putting the chamomile and all the spices into a square of muslin and tying the ends tightly with string, a bit like a bouquet garni. Then put into a saucepan with the milk and sugar. Using a slow heat, bring to the boil and cover. Simmer for 10 minutes. Allow to cool to drinking temperature and then remove the muslin bundle.

Chamomile Panna Cotta

You can use the same method of infusion as in the previous recipe, using only chamomile in the muslin bundle, to make this chamomile panna cotta. Your guests will immediately recognise the delicate flavour as soon as you tell them, but you should be able to keep them guessing for a while.

SERVES 6

70ml skimmed milk
2 tsp agar-agar flakes or equivalent amount of powder
600ml double cream
113g granulated white sugar
30g dried chamomile flowers, tied into a muslin bundle

Pour the milk into a small bowl, and stir in the agar-agar flakes or powder. Set aside.

In a saucepan, stir together the cream and sugar and cook over a medium heat. Add the bundle of chamomile flowers and bring to a full boil, watching carefully, as the cream will quickly rise to the top of the pan. Pour the agar-agar and milk into the cream, stirring until completely dissolved. Cook for 1 minute, stirring constantly. Remove from the heat and pour into 6 individual ramekin dishes.

Leave to cool at room temperature, uncovered, so that the steam doesn't make the mixture soggy. When cool, cover with clingfilm, and refrigerate for at least 4 hours, but preferably overnight, before serving. You might like to dish up the panna cotta with some fresh chamomile flowers – including their stems – on the side of the plate.

You can replace the chamomile flowers in this recipe with pineapple weed, elderflower or meadowsweet, to make different-flavoured panna cottas.

(Sweet) Chestnut

Castanea Sativa

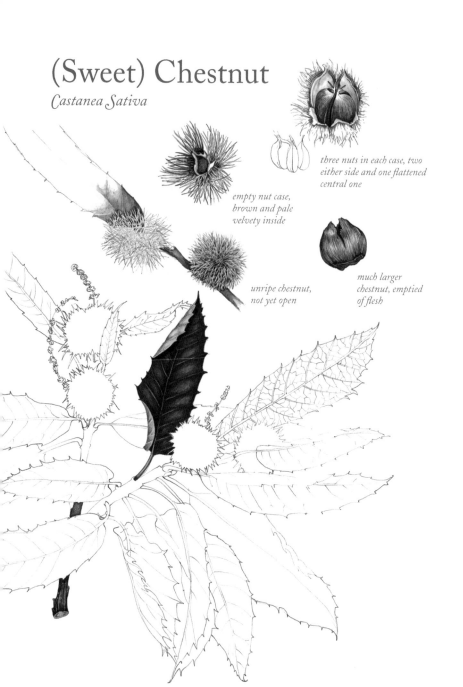

three nuts in each case, two
either side and one flattened
central one

empty nut case,
brown and pale
velvety inside

much larger
chestnut, emptied
of flesh

unripe chestnut,
not yet open

The tree that gives us delicious sweet chestnuts grows to about 30 metres. Its bark is thick and brown, with spiral-shaped cracks, and the leaves are longer than they are wide, with jagged edges, dark green on the top and a lighter green underneath. The catkins that start to appear in early spring and on into the summer are 12–20cm long, and the nuts appear shortly afterwards.

How to tell the difference between chestnuts and conkers

This is a moot point. Conkers (the nuts from the horse chestnut tree) are not edible, so take care not to confuse them with chestnuts. If you eat a conker by accident it won't kill you, but it won't make you very happy either. Luckily it's quite easy to tell them apart. The outer shell of the conker has lumpy, pointy bits, like a green sputnik, and is thick and coarse. The shell of the sweet chestnut, on the other hand, is prickly, with finer points. Conkers tend to be bigger, too. Conkers are sometimes boiled and used in animal feed, but in general I'd say it's best to use them in the traditional manner – i.e. attach a couple of them to shoelaces and then bash the heck out of each other!

Culinary uses

It's a real shame that we tend to consign sweet chestnuts to Christmas and Thanksgiving dishes. In mainland Europe they're used far more extensively, and they're so delicious that I think we should follow suit. The *Castanea sativa* tree originated in Greece, but there's also an American version, *Castanea dentata*. Chestnuts contain almost twice as much starch as the potato and, allegedly, the Greek army survived on them during their arduous retreat from Asia Minor in 401–399 BC.

Because of their starchy content, sweet chestnuts can be ground to a fine meal and used in bread-making, hence the tree is sometimes called the 'bread tree'. However, if you decide that you love these nuts enough to want to plant a tree in your garden you'll have a long wait until you can roast them on your fire. It takes at least 40 years for the tree to produce nuts. Your children, grandchildren, great-grandchildren and even your great-great-great grandchildren will thank you for it, though, as the trees can live for 500 years or more.

How to cook chestnuts

By far the most common way to cook chestnuts is to simply roast them in their shells. Score a cross in the nuts with a very sharp knife and put them in a hot oven for 30 minutes or so – or even better, on top of a hot wood burner, or on a shovel in an open fire (this last method needs careful attention if you don't want your carefully-gathered chestnuts to turn into cinders). Then you pop them out of the shells and eat. If you want to get fancy, you can dip them in butter and salt.

Alternatively, score the shells as above, then cover with water and simmer in a pan for 15–20 minutes or so until the nuts are soft. Leave to cool completely and you'll find that with any luck the shells and the furry inner lining come away easily from the chestnut. To make, say, 240g chestnuts you'll need approximately 300g raw nuts. This should account for any 'accidents' and also for the ones that don't make it into the recipe, but end up in your mouth instead.

Warning: Don't eat chestnuts in their raw state. The skins contain high levels of tannic acid and will upset your stomach.

Medicinal uses
Before inoculation against whooping cough was introduced, fresh sweet chestnut leaves were used to soothe the symptoms of this childhood ailment.

Did you know?
Unlike other nuts, chestnuts have a low fat content. Even more reason to use them!

Chestnut Brownies

Here's a recipe for some yummy brownies. Make sure you don't let them cook *too* long, as you want your brownies to have a lovely hot doughy centre.

CUTS INTO 16 X 4CM SQUARES

240g shelled cooked chestnuts
200g brown sugar
2 tsp vanilla extract
25ml dark rum
100g plain flour
2 medium eggs, separated
200g unsalted butter
200g dark chocolate, broken up

Preheat oven to 170°C/gas mark 3. Line a deep, 20cm square baking tin with baking parchment. Chop the cooked chestnuts, put them into a bowl, add half the sugar, the vanilla and the rum and mix well. Fold in the flour gradually.

In a separate bowl, whisk the egg whites until they form soft white peaks. Gently stir in the other half of the sugar to make a soft meringue mixture, then beat in the egg yolks.

Melt the butter and chocolate in a heatproof bowl suspended over a pan of simmering water, remove from the heat and beat into the flour and chestnut mixture. Stir this mixture into the meringue and then spoon the whole thing into the tin. Bake for 20–25 minutes, until the mixture is only just set in the middle. This will ensure the brownies are nice and squidgy and gooey … yum! Leave to cool completely before cutting.

Sweet Chestnut and Ricotta Cheesecake

MAKES 1 X 25CM CHEESECAKE

500g fresh chestnuts, shelled (this will mean you need 600-700g unshelled chestnuts)
1 vanilla pod, split lengthwise
60ml dark rum
100g granulated white sugar
1–2 tbsp water, if needed

FILLING
1kg good-quality ricotta cheese
100g granulated white sugar
4 large eggs, beaten
2 tsp vanilla extract
4 tbsp candied lemon peel, finely chopped

First, score the chestnuts and roast in a hot oven (220°C/gas mark 7) or simmer in a pan before peeling (see pages 51–2) This should give you about 500g shelled nuts.

Turn the oven down to 180°C/gas mark 4.

Mix the chestnuts in a large, heavy-bottomed pan with the vanilla pod, rum and sugar. Cook fast on a high heat for 5–8 minutes, stirring so the mixture doesn't burn. Add a little water if you need to, but bear in mind that you're aiming for a thick consistency. Allow to cool, then crush into small pieces.

While the chestnut mixture is cooling, grease a 25cm springform cake tin. Mix together the ricotta, sugar, eggs and vanilla extract. Fold in the cooled chestnut mixture, making lovely marbled streaks in the ricotta. Scoop into the cake tin and bake for 1 hour, then turn the oven up to 190°C/gas mark 5 and bake for a further 15 minutes. When cooked, a knife inserted into the cheesecake 5cm from the edge should come out clean. Let the cake cool, then run a small palette knife round the edge to loosen the cheesecake before removing it from the tin.

Sticky Sweet Soy Chestnuts

30 fresh chestnuts
380g granulated white sugar
50g packed dark brown sugar
120ml soy sauce
120ml water
1 tbsp olive oil
¼ tsp salt, plus more for sprinkling

Roast the chestnuts (see pages 51–2). Remove the shells, keeping the chestnuts as whole as you possibly can.

While the nuts are roasting, make the sauce. Combine the sugars and the soy sauce in a heavy-bottomed pan over a high heat, mixing until dissolved. Leave for 10–12 minutes until the mixture is bubbling. Add the water, stir well, and simmer for another 10 minutes or so. Remove from the heat, stir in the oil and salt. Toss the chestnuts in the sauce, covering well. Allow to cool. Eat!

Traditional Marrons Glacés

Forget the scent of coffee and freshly baked bread – if you want to sell your house, then make sure you have some marrons glacés in the oven when prospective buyers come to visit. Marrons glacés are a traditional treat in mainland Europe and I see no reason why they should have all the fun. They're not hard to make and the only problem will be trying not to eat them all yourself.

MAKES APPROX. 1KG

1kg chestnuts, shelled
600ml water
1kg granulated white sugar
1 tsp vanilla extract

Place the chestnuts in a large pan with just enough water to cover. Bring to boil and cook for 10 minutes. Drain the chestnuts and discard the cooking liquid. Using a clean tea towel (some people prefer to use their fingers), rub the thin skin off the cooked chestnuts.

In a separate pan, bring the water, sugar and vanilla to a boil, stirring constantly. Continue cooking, stirring occasionally, for 5 minutes.

Add the prepared chestnuts to the boiling sugar syrup and stir the chestnuts until the whole mixture returns to a boil. Continue cooking the chestnuts, stirring frequently, for 10 minutes. Now pour the candied chestnuts, along with the sugar syrup, into a large, heatproof bowl or container, and loosely cover it. Allow the chestnuts to soak in the syrup for 12–18 hours.

Add the chestnuts and syrup to a clean pan and repeat the process, this time boiling them for 2 minutes, and then soaking the mixture, loosely covered, for 18–24 hours. Repeat the entire process 3–4 times in total, until the sugar syrup has been absorbed by the chestnuts.

Preheat oven to 130°C/gas mark ½ and arrange the candied chestnuts on a baking tray lined with baking parchment. Place in the oven and turn off the heat. Allow the chestnuts to dry in the oven for 45 minutes to 1 hour, until they have firmed up and the surfaces of the nuts are dry. Transfer to presentation boxes lined with greaseproof paper for the perfect gift. Otherwise, store in an airtight container.

Chicory

Cicorium Intybus

single flower

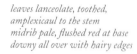

*glorious cerulean/cobalt blue petals,
stamens white, anthers dark blue*

*leaves lanceolate, toothed,
amplexicaul to the stem
midrib pale, flushed red at base
downy all over with hairy edges*

The tall stems (30–100cm) of the chicory plant are dotted with distinctively pretty blue (or sometimes white or lilac) flowers during the months of July to October. The stems are tough and bristly and the leaves are rectangular, with bristles on their undersides. The plants are perennial, so if you find some make a note of where they are for the next year. Chicory likes to grow in scrubby land and so can often be found in roadside verges.

You might find that chicory is sometimes referred to as 'endive'; however, the endive is actually a different plant altogether.

Culinary uses

If you've ever eaten the cultivated chicory that's available in supermarkets you'll be familiar with the bitter tang. It's an acquired taste, but once you've acquired it, you'll appreciate the distinction it gives to certain dishes. You can blanch chicory and use it as you might spinach, to add flavour, colour and bulk to pasta dishes. Chicory sautéed with garlic and shallots, or garlic and red pepper and a touch of chilli, is particularly delicious and makes a good side dish or tapas-style delicacy.

One of the common names of chicory is 'coffee weed' and this gives us a big fat clue as to another use of the plant. It's been used both as a

coffee substitute – although really it has a taste of its own that certainly isn't like coffee – or as an addition to coffee powder. It seems to have come into its own during times of conflict; for example, it was used during the Napoleonic Wars when coffee was scarce and expensive, and again in the Second World War. And you can still buy it over the counter in liquid form; otherwise, have a go at making it yourself.

Medicinal uses

Chicory is said to protect the liver from the effects of excess coffee consumption so it's very appropriate that it is often blended with coffee or used as a substitute for it. Otherwise, it has a tonic effect on the liver and the digestive tract and can be used both as a laxative and a diuretic.

Did you know?

Chicory is one of the first plants to have been mentioned in literature. The Roman poet Horace names it in relation to his own simple diet, and includes olives and mallows with it in his list.

Chicory 'Coffee'

Having spotted your chicory plants during the summer, head back to them in the autumn. It's the roots you're after, and they'll be fatter now and ready to harvest. The tricky bit is digging them up without damaging them or breaking them up too much, especially if they're growing in compacted, scrubby ground. It's easier in wettish soil. Dig around them, going deeper than you might think; the roots can be up to 76cm long. So, heave up your bunch of roots, then wash them thoroughly in clean running water with a scrubbing brush. Slice the roots into discs of about 2–3mm thick.

Next, you need to dry them. If the weather is warm, then you can do this outside. If you've an Aga-style oven, spread the roots on top of it. If not, set your oven on its lowest heat and let the roots dry out gradually on a lined baking tray. The roots should feel slightly brittle to the touch and will look a bit like dried mushrooms.

Leave the chicory roots to cool before you roast them. Again, how much you want to roast them is up to you, but colour is a good gauge. In a slow Aga, for example, roast them for about an hour to get the good dark brown colour. A 'normal' oven will take up to three hours, depending on its behaviour; they're all different. A lovely side effect is that the roasting roots smell lovely; like a rich, chocolatey coffee scent.

Grind the roots in a coffee grinder. They won't grind down as evenly as commercial coffee beans but this doesn't matter.

Make the chicory drink in exactly the same way that you'd make coffee in a cafetière with boiling water. Allow it to steep for a couple of minutes, add sugar and milk to taste – *voilà*. You can add the powder to milk and sugar and simmer to make a sort of chai, or use it to flavour desserts and cakes.

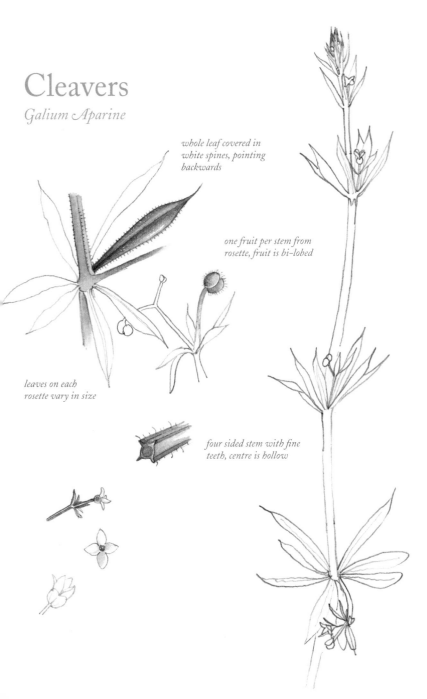

Cleavers
Galium Aparine

whole leaf covered in
white spines, pointing
backwards

one fruit per stem from
rosette, fruit is bi-lobed

leaves on each
rosette vary in size

four sided stem with fine
teeth, centre is hollow

Cleavers are those long, whip-like plants with sticky stalks and hundreds of teeny sticky balls, which are actually their seeds. You might know them better as 'goosegrass'. Children use the stalks to stick on each other's back and dogs frequently get covered in them! Those stems can grow up to 120cm, and the leaves are borne along the stems in circular designs.

Culinary uses

Don't bother with the stems of cleavers – they're not only tough but don't taste great either. But the leaves, which are at their best when they are young, are lovely and taste a bit like pea shoots. You can use them as you would any other leaf vegetable – steamed, sautéed, or eat them raw in salads, etc.

And, those sticky seeds, so abundant, make a tasty drink if you dry and roast them and then grind them in a coffee grinder. Then make just like coffee, in a cafetière.

Medicinal uses

Used externally, cleavers can ease ulcers and wounds. Taken internally they can alleviate the pain of cystitis. Cleavers can also help to bring down a high temperature.

Did you know?
The old Greek name for cleavers is *Philanthropon*, meaning 'love', because of the clinging habit of the plant.

Cleavers Curry

This is a very unusual – and utterly delicious – recipe for a curry, using fresh cleavers leaves. It's great mopped up with fresh crusty bread or naan, instead of rice.

SERVES 2

2 onions
2 garlic cloves
Oil, for frying
Piece of fresh ginger, chopped
3 tsp (mixed) fresh coriander, cumin
  powder and chilli powder
2 carrots
2 tomatoes
3 handfuls young cleavers' leaves
250ml water
100g creamed coconut milk

Cut the onions and garlic and fry in some oil on a medium heat. Stir in the chopped ginger and all the spices, stirring all the time. Chop and add the carrots, tomatoes, and cleavers leaves. Add the water and the coconut milk and simmer for approximately 15 minutes until cooked. Serve with rice.

Cleavers Soup

Another interesting recipe that's worth a try, but make sure that you only use fresh, young cleavers leaves. See if your guests can guess what you've served them!

SERVES 4

3 handfuls young cleavers leaves
2 handfuls mixed leaves (e.g. ribwort
  plantain, nettle, dandelion, yarrow,
  clover, daisies)
Herbs, for seasoning (e.g. lovage, thyme,
  marjoram, lemon balm, peppermint,
  dill, parsley, chives)
1 onion
Oil, for frying
2 garlic cloves
1 potato, diced into small pieces
750ml vegetable stock

Wash the cleavers leaves, weeds and herbs and chop them coarsely.
   Chop the onion and the garlic and fry in oil until golden. Add the potatoes, cleavers leaves, mixed leaves and herbs. Pour in the stock and cook for about 10 minutes. Let cool slightly, then liquidise until the soup is smooth.

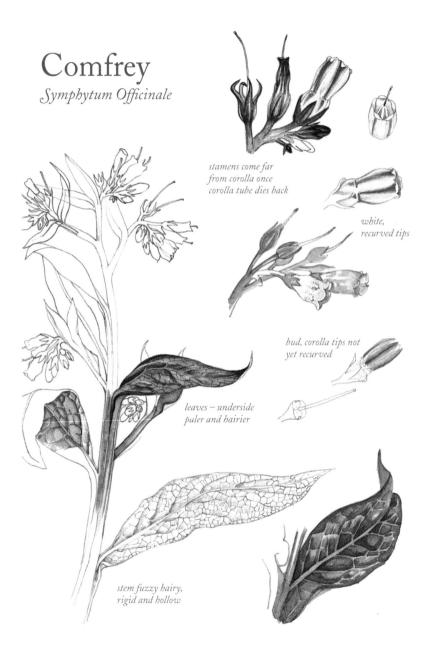

Comfrey

Symphytum Officinale

stamens come far
from corolla once
corolla tube dies back

white,
recurved tips

bud, corolla tips not
yet recurved

leaves – underside
puler and hairier

stem fuzzy hairy,
rigid and hollow

Comfrey likes to grow in shady, damp places such as woodland riverbanks and of course the hedgerow. It's easy to spot since it's quite tall (approx. 1–1.2 metres high) and has large leaves (the lower ones can be especially big, up to 30cm). Comfrey is a fuzzy, hairy plant and its flowers can be green-white, pink, or purplish. It looks a little similar to the foxglove, but if in doubt refer to the flowers; foxgloves have graduated bell-like flowers spaced regularly all the way up the stem, whereas comfrey blossoms appear in dangling clusters.

Culinary uses
Gather the young leaves – the older ones are a bit tough and bitter – and use in the same way as spinach. The stalks can be blanched and eaten with butter and a little lemon juice. The roots, too, can be dried and roasted in the same way as chicory or dandelion root, and used as a drink.

A word of warning, however. As recently as the 1970s comfrey was hailed as a kind of vegetarian superfood. However, the recent discovery of its high pyrrolizidine-alkaloid (PA) content, including the carcinogen symphytine means that the consumption of the leaves and especially the root as a tea in large quantities is not recommended nor advisable, and pregnant women, children and anyone suffering from a chronic disease or taking medication should avoid comfrey altogether. If you are in doubt, it's probably best to do likewise or ask for the advice of your GP.

Did you know?
Comfrey makes a great plant food. Simply add put a few handfuls in a bucket, top up with water and leave to leach for a few days or until the water goes brown and smelly, then use to water your plants, in particular tomato and French and runner bean plants which thrive on it.

Comfrey and Mushroom Parcels

SERVES 4 AS A MAIN COURSE OR 6 AS
A STARTER

4 onions, chopped
100g butter
225g mushrooms (cultivated or, if you
  know your mushrooms, wild)
450g cooked comfrey (you'll need twice
  this amount before cooking since, like
  spinach, the leaves reduce down
  significantly)
300ml white sauce
Dash of soy sauce
2–3 garlic cloves, crushed
A handful of chopped herbs
450g puff pastry (frozen sheets are fine),
  defrosted
1 egg, beaten

Preheat oven to 200°C/gas mark 6.

In a heavy-bottomed pan, sauté the onions in the butter, until golden. Add the garlic and mushrooms and continue cooking until they're soft but not mushy.

Combine the comfrey with the white sauce, drain the onion and mushrooms and stir into the comfrey. Add the chopped herbs, a dash of soy sauce and season to taste.

Roll out the pastry to about 3–4mm thick, place on a greased baking tray and dampen the edges with water. Spoon the mixture down the centre of the pastry, leaving a decent border along the edges, then fold over the edges to make a long parcel. Prick the top with a fork and brush with the beaten egg.

Bake in the oven for about 25 minutes, until the parcel is risen, golden and lovely.

Comfrey Fritters

SERVES 4 AS NIBBLES

100g plain flour
1 egg, separated
600ml lukewarm water
Oil, for frying
Handful of young comfrey leaves,
  washed
Salt and pepper, to taste

To make the batter, sieve the flour
into a bowl and drop in the egg yolk.
Gradually add the water and beat
well with a spoon. In a separate
bowl, whisk the egg white until it
forms stiff peaks, then fold it into
the batter mix.

Heat the oil in a deep, heavy-
bottomed pan or a deep-fat fryer.
When the oil is hot, dip the comfrey
leaves into the batter and deep-fry
them in the hot oil. Remove with
a slotted spoon and drain on kitchen
paper. Serve sprinkled with salt
and pepper.

Crab Apple
Malus Sylvestris

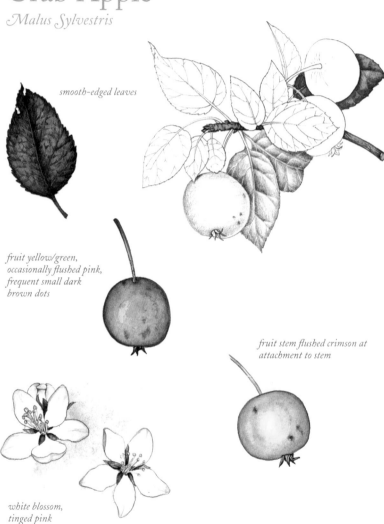

smooth-edged leaves

*fruit yellow/green,
occasionally flushed pink,
frequent small dark
brown dots*

*fruit stem flushed crimson at
attachment to stem*

*white blossom,
tinged pink*

Crab apple trees are found all over the UK, both in hedgerows and on open land. The trees can grow to 10 metres, and the older the tree, the more apples it yields.

Crab apples are much smaller than 'normal' apples, fairy-like versions of their larger cousins – about 2cm in diameter, yellow in colour, sometimes with a reddish blush. If you've ever bitten into a crab apple, thinking it would taste sweet, you'll know the shock of the crab apple's knife-like bitterness. The acidity of the crab apple has been known for millennia; there's evidence that our ancestors enjoyed cider made from the fruit, 4,000 years ago.

Culinary uses

The fruits of the crab apple tree are prolific. But because they're so bitter you can't simply use them as a substitute for regular apples – try recipes such as crab apple jelly and crab apple cheese.

Did you know?

Apple trees – including the crab apple – are considered to be amongst the most magical of trees. Some believe that if you fall asleep under one, you'll be fair game for the fairies to carry you off.

When you call someone 'crabby', meaning bad-tempered, you're referring to the bitterness of the fruit!

Crab Apple Jelly

MAKES 4 X 454G JARS

2kg crab apples
Approx. 1.25 litres water
Granulated white sugar (the quantity depends on how much juice the apples yield)

Wash and dry the apples and chop roughly. Put in a large, heavy-bottomed pan and cover with water. Slowly bring to the boil and simmer for 1 hour, stirring from time to time to help break the apples down.

Ladle the apples into a fine jelly bag and allow the juice to drip into a pan, ideally overnight. Measure the juice, then add 450g sugar for every 600ml juice. Place the juice and sugar back in the pan, bring to the boil and leave on a fast rolling boil until it reaches setting point, which will take about 15 minutes. This is best tested by dropping a blob of the mixture onto a chilled saucer – the blob should wrinkle on the top if you push it with your finger.

Remove from the heat, skim any scum from the surface, and pot into warmed, sterilised jars.

Crab Apple Cheese

MAKES APPROX. 1KG

1.5kg crab apples
300ml sweet cider
300ml water
Soft brown sugar
½ tsp each powdered cinnamon, cloves
  and nutmeg

Wash and dry the apples and chop roughly. Put into a pan with the cider and water and simmer until soft. Press the solids through a sieve, weigh, and add 450g sugar to every 600ml of purée.

Put the apples and sugar into a heavy-bottomed pan and stir over a low heat until the sugar is dissolved. Add the spices, bring the mixture to the boil, then reduce the heat and simmer until the mixture has thickened. Keep stirring and don't let it burn. Pour into warmed, sterilised jars, and seal.

This is great with strong cheese and crackers, or spooned over ice cream.

Crab Apple Cake

Now for something more unusual … This cake has a delicious sharp tang that's offset beautifully by crème fraîche or ice cream. This recipe and the spicy crab apple marmalade that follows are courtesy of Pille Petersoo.

SERVES 4

PASTRY
200g unsalted butter
200g sour cream
350g plain flour
½ tsp salt
1 egg white, for brushing
Icing sugar, for dusting

FILLING
Approx. 400g crab apples, cleaned weight
Granulated white sugar, to taste (start
  with 4 tbsp)
Cinnamon, to taste

Melt the butter and, in a bowl, mix with the sour cream, flour and salt until combined. The dough should be very soft at this stage. Cover the bowl with clingfilm and place in the fridge for half an hour.

Quarter and core the crab apples. Place in a bowl and cover with water to stop them from browning. Preheat oven to 200°C/gas mark 6. Line a baking tray with greaseproof paper.

On a lightly floured surface, roll the chilled dough into a large rectangle. Carefully lift the dough and lay half of it on the baking tray, letting the other half hang over the edge (it will fold over later to make the crust).

Spread the apples over the pastry on the tray, then season with sugar and cinnamon. Fold over the other half of the pastry and press the edges firmly together. Brush with egg white and sprinkle with more sugar.

Bake in the middle of the oven for about 35–40 minutes, until the apples are cooked and the cake is a lovely golden brown on top.

Remove from the oven and leave to cool a little before dusting with icing sugar.

Cut into squares to serve.

Spicy Crab Apple Marmalade

MAKES APPROX. 2 X 454G JARS

1kg crab apples
200ml water
2 cloves
Good knob of fresh ginger, chopped
1 sweet chilli pepper, de-seeded and
  chopped
1 cinnamon stick
250g caster sugar

Wash the apples and place in a saucepan with the water, cloves, ginger, chilli and cinnamon. Bring to the boil and simmer on low heat, stirring every now and then, until the apples are soft.

Using a wooden spoon, press the softened apples through a sieve (try to get as much of the apple fruit and peel through as possible).

Return to the saucepan, add the sugar and bring to the boil. Simmer gently for another 7–10 minutes. Spoon the jam into warmed, sterilised jars, cover with wax discs and seal. Store in a cool place.

Damson
Prunus Domestica

stems of leaves flushed
crimson

fruit covered in
bloom, slight cracks
have orange tint,
mottled bloom with
darker spots

leaves have rounded tops,
double serration on edges

purplish bark flushed
orange and with pale dots

The damson is the little sister of the plum (indeed, it's a 'wild' plum) and the big sister of the sloe. You might see the tree tucked in amongst old hedgerow shrubs, at the edges of fields or in native woodland. It's similar to the blackthorn (which is where the sloe comes from) but without the vicious thorns. Damsons are smaller fruits than plums and a blackish-purple colour.

Culinary uses

Unlike sloes, damsons are sweet and can be eaten directly from the tree without it feeling like your mouth is being turned inside out!

Medicinal uses

Plums are rich in vitamin C, which can help boost the immune system. They also contain sorbitol, which helps to regulate the digestive system, hence their use as a laxative.

Did you know?

The first recorded mention of the damson was 2,000 years ago and the plant itself could have originated in Damascus, which is why it was called the plum of Damascus, from which the damson gets its name.

The Balkan spirit, slivovitz, is made from distilled damson juice.

Wild Damson Chutney

MAKES APPROX. 2 X 454G JARS

450g damsons, washed, dried and stalks removed
1 large baking apple, peeled, cored and chopped
1 large onion, chopped
75g sultanas
150ml (¼ pint) vinegar
1 tsp salt
Pinch of mustard powder
7g whole cloves and fresh ginger
½ dried chilli pepper (or a whole one if you like a serious kick)
350g granulated white sugar

Simmer the damsons in water until they're soft, but still relatively intact. There's no need to peel or stone, since the skin and stones will fall away easily and all you need to do is lift them out with a slotted spoon.

Put the fruit into a thick-bottomed pan with all the other ingredients except the sugar, tying the spices up in a muslin bag like a bouquet garni. Simmer for about half an hour, then add the sugar and bring to a rolling boil, stirring from time to time and making sure you don't burn the mixture. Allow to cool, then pour into warmed, sterilised jars. Cover and seal.

Damson Gin or Vodka

Sloe gin might be better known, but damson gin – or vodka – is a twist on an old favourite. It's ridiculously easy to make. You'll need a large, sealable jar, big enough to take all the ingredients.

MAKES I X 70CL BOTTLE

450g damsons, washed, dried and stalks
*   removed*
200g granulated white sugar
70cl bottle of your favourite gin (it's a
*   false economy to use nasty gin just*
*   because you're adding other*
*   ingredients to it)*

Freeze the damsons for a couple of days – this helps break down the skin and allows the juices to penetrate the gin better. And it's quicker than painstakingly pricking the fruits with a pin.

Put all the ingredients into the jar, seal and leave in a dark place, for up to 6 months, shaking the jar occasionally, then strain the alcohol from the fruit. It should be a lovely dark pink colour. If you leave the fruit in too long, then the resulting liquid won't be such a pretty colour.

Damsons in Syrup

This is a very easy and delicious way of preserving a glut of damsons. A few aniseed seeds added to the preserving syrup imparts a lovely spicy flavour. If you have a few jars of these, then you can make sugar plums (or rather, sugared damsons) next.

600ml water
250g sugar
A few aniseed seeds (optional)
500g damsons, washed, dried, and
*   stalks removed*

Simply put the water, sugar and aniseed seeds, if using, into a large pan and heat until the sugar has dissolved. Boil the syrup for a minute or so. Leave to cool.

Put the damsons into a large, sterilised jar and pour the cooled syrup over the top. Seal the jar tightly. You can preserve other fruits in this way, too.

Sugared Damsons

This recipe might sound time-consuming but it isn't really; the plums need to spend a long time in a cool oven, otherwise your part in the process is pretty minimal. If you happen to have a range oven or an Aga, so much the better, but a normal oven on its lowest setting will do nicely. Sugared damsons were a Victorian delicacy, particularly popular at Christmas, and a treat that really ought to be revived.

Once your damsons have been sitting in the syrup for about 3 months (see previous recipe), they are ready to be transformed.

You will need, per 500g jar of plums, the same amount of white caster sugar.

Line a baking tray with greaseproof paper. Taking each plum in turn, shake off the excess syrup and roll in the sugar. Put aside for an hour or so, then repeat the process. Lay the fruits on the tray and leave to dry in the cool oven for several hours. Then re-coat the plums, set aside again, coat again, and bake again, using a clean sheet of baking paper.

The whole process needs to be repeated 4 times and can take as many days before the plums are dry and covered in a crisp sugar coating. When cool, place in boxes to make a very charming gift.

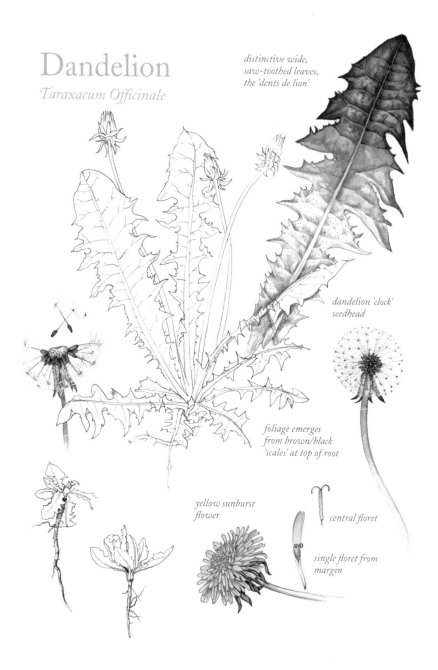

Dandelion
Taraxacum Officinale

distinctive wide,
saw-toothed leaves,
the 'dents de lion'

dandelion 'clock'
seedhead

foliage emerges
from brown/black
'scales' at top of root

yellow sunburst
flower

central floret

single floret from
margin

Is there really any need to tell you exactly what a dandelion looks like? It is of course the lovely bright yellow sunburst of a flower that proliferates through spring until the end of summer, when its dandelion 'clocks' cast soft seed heads into the breeze. The leaves are very distinctive, too, with their wide, saw-tooth edges. The stems are hollow and smooth, with a milky sap inside that oozes out if you break them open.

Culinary uses

Dandelion leaves make a great salad vegetable as long as the leaves are young; beyond the soft pale green stage they tend to be bitter. If you start to really get into dandelion leaves, you can put terracotta plant pots over the young shoots. This not only blanches them, but allows them to grow taller as they reach for the light coming through the holes in the bottom of the pot. The older leaves can be used in the same way as spinach; for example, wilted or cooked down with garlic and a squeeze of lemon. Mrs Beeton advises steeping them in cold water for an hour or so before cooking with butter. Try them cooked in hot sesame oil and sprinkled with toasted sesame seeds to add an oriental flavour.

You can make dandelion tea very easily; just put a handful of washed leaves in a teapot, add boiling water, steep for 5 minutes and add sweetening if you want.

Dandelion flowers are also edible. The bright colour of the petals make a good addition to salads, pancakes and even jellies and trifles. You can also eat the buds – try them thrown into fresh egg pasta with olive oil and a little salt and pepper.

Medicinal uses

Dandelion is a natural diuretic and dandelion tea is therefore said to be an effective remedy for water retention and cellulite. This same diuretic action helps flush out toxins, including uric acid, alleviating rheumatism and gout. Dandelion also helps to detoxify the liver.

Did you know?

The name 'dandelion' comes from the old French name for the plant *dent-de-lion*, meaning 'lion's tooth', referring to its jagged leaves. There's another, more recent French name that refers to its diuretic qualities; *pissenlit*, meaning 'wet the bed'.

You can make a rubber band out of a dandelion – break the stem of the plant and spread the oozy sap thinly along your finger. Wait for it to dry, then peel off the ends and tie together … hey presto. This same milky sap is said to get rid of warts.

Dandelion 'Coffee'

This is surprisingly easy to make and delicious to drink. All you need are the roots of the plant and a little time. Incidentally, this doesn't taste like coffee at all but it's hard to know what else to call it.

Choose dandelions that are growing in the wild because the roots will probably be bigger and fatter than those growing in a garden. Also, the roots will be at their biggest and best in the autumn and winter months after the plant has spent the summer months storing up energy reserves.

Dig up the roots when the soil is wet and they'll be easier to get at. Wash thoroughly (both the roots and yourself) and then leave the roots for a couple of days in a warm, dry place – an airing cupboard, for example. Slice into 2cm chunks and then chop roughly in a food processor.

Preheat oven to 200°C/gas mark 6, spread the roots on a lined baking tray, and roast for a couple of hours with the oven door open to allow any moisture to escape. You'll need to turn the roots from time to time to ensure that they cook evenly. You might also want to experiment with roasting times; this is a matter of personal taste, just as with coffee beans.

You can then grind the roots finely in a coffee or spice grinder and either use straight away or store in a jar.

To make a mug of the 'coffee', simply add boiling water and sugar (if you want) to a heaped teaspoon of the ground dandelion root.

Dandelion Flower Wine

For this recipe you will need a large plastic food-grade container, a demijohn with airlock and some sterilised bottles with corks.

MAKES APPROX. 3 X 70CL BOTTLES

1 unwaxed orange
1 unwaxed lemon
3 litres in volume of fresh young
  dandelion flowers, washed and dried
4.5 litres water
1.25kg granulated white sugar
One 7g sachet of wine yeast, made
  according to manufacturer's
  instructions

Peel the fruit, removing as much of the pith as possible, and put the peel with the dandelion flowers into the centre of a square of muslin. Tie the end of the parcel with string and boil in the water for about 15–20 minutes. Remove the bag, squeeze it over the pan and add the sugar to the liquid. When the sugar has dissolved, pour the liquid into a large plastic food-grade container and add the juice of the fruit. Once the liquid has cooled to blood temperature, add the yeast, as per the manufacturer's instructions.

Leave, covered, for 3 days in a warm place to allow the fermentation process to start. Then siphon into a demijohn, being careful not to disturb any sediment at the bottom. Put an airlock in the neck of the demijohn jar and leave in a warm, dark place. Once the bubbles have stopped popping up in the fermentation lock, siphon the wine into sterilised bottles, and cork. Leave to mature for about a year if you can bear it!

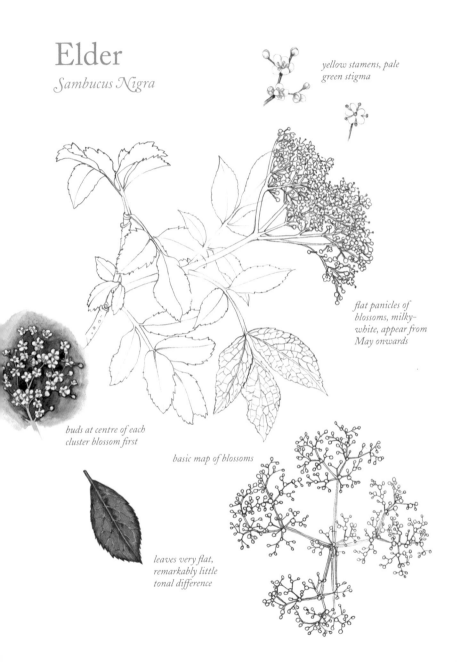

Elder

Sambucus Nigra

yellow stamens, pale green stigma

flat panicles of blossoms, milky-white, appear from May onwards

buds at centre of each cluster blossom first

basic map of blossoms

leaves very flat, remarkably little tonal difference

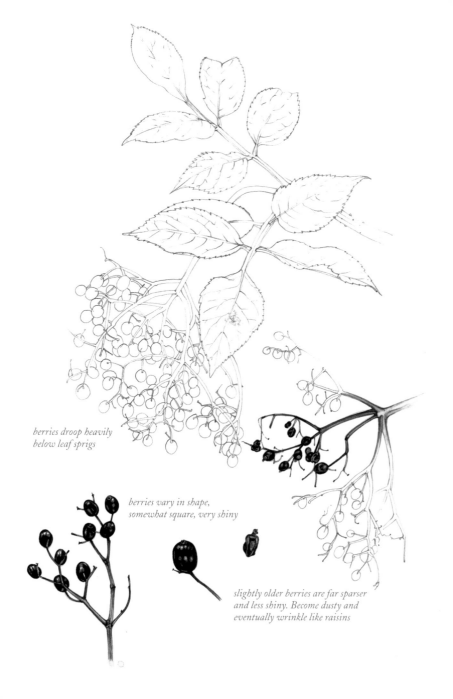

*berries droop heavily
below leaf sprigs*

*berries vary in shape,
somewhat square, very shiny*

*slightly older berries are far sparser
and less shiny. Become dusty and
eventually wrinkle like raisins*

Elder is a smallish to medium-sized deciduous tree or shrub, with oval, saw-toothed leaves, 5 or 7 to a stalk. It can grow up to 10 metres tall. The bark is a browny-grey colour, smooth, often with a coating of green lichen. It's seen in hedgerows and in dappled woody areas as well as on its own. The flat frothy panicles of flowers, made of thousands of tiny blossoms, appear from May onwards, and the blackish-purple berries follow from August.

Culinary uses
The exciting thing about the elder is the sheer versatility of its flowers and berries. The elder is used in the liqueur 'sambuca' – so-called because of the plant's Latin name.

Medicinal uses
Elderflowers have a soothing effect and are sometimes used in eye lotions. The berries have a high vitamin C content and have been shown to be highly effective in treating colds and flu. Gargling with a decoction of the leaves and stems will soothe a sore throat. The flowers, leaves, stem and root of the elder have laxative and diuretic qualities. Now, although I never intended this to be a book of 'remedies', I have broken my own rule here and included the following recipe, which not only tastes great but is the best flu remedy I have ever come across. Thanks to Heathcliffe Bird for bringing it to my attention (see page 88).

Did you know?
The old Anglo Saxon name for elder is 'aeld', meaning fire. This is because the branches are hollow and were once used like bellows, to blow air through to help fires to start. These hollow branches can also be used as pea-shooters.

ELDERFLOWERS

If there's one scent that is the epitome of an English summer, then it has to be the fragrantly heady scent of elderflower.

If you look closely at the frothy, creamy white umbels you'll see that they're actually composed of a loose explosion of thousands of tiny flowers, and it's the pollen of these flowers that produces that unmistakably beautiful aroma (I often wonder why the elderflower doesn't seem to have been explored very much by perfumiers).

The best way to capture that summer scent and fragrant flavour is in an elderflower cordial, which forms the basis for many recipes.

Tips for the best elderflower taste
Be sure to pick the flowers when the heat of the sun is on them. And make sure that there's plenty of pollen on the blossoms.

Don't bother gathering flowers immediately after rain; it's the pollen that gives the flavour, and the rain will wash it away.

For the same reason, don't wash the flowers. Any insects or debris will be strained away later in any case.

Basic Elderflower Cordial

This is so easy to make, it's a good idea to make as much as you can, then pour into plastic bottles and freeze, as there are so many things you can use it for. And after all, the season for elderflowers is short.

MAKES APPROX. 1.5 LITRES

1.5kg granulated white sugar
1.7 litres boiling water
20 heads elderflower
50g citric acid
2 unwaxed lemons, sliced

Dissolve the sugar in the boiling water and leave to cool. Once the sugar solution has cooled, put all the ingredients in a bowl, cover, and leave for 24 hours at room temperature. Then strain and pour into bottles. You might also want to freeze the cordial into ice cubes and use to flavour cocktails – for example, a couple of chunks thrown into a tall glass with a mix of decent vodka and sparkling water, and maybe a slice of lemon, comes highly recommended!

As a variation on this recipe, try substituting limes for lemons.

Elderflower Lemonade

This is sometimes called elderflower champagne. The first time I drank this was from a batch made by a wonderful man called Colin Fox, for the wedding of his daughter Natasha to Adrian. Colin even drew beautiful Celtic knotwork labels for the lemonade.

MAKES APPROX. 4.5 LITRES

2 litres in volume of elderflowers, after
the stalks have been removed
2 litres water
1 unwaxed lemon, sliced
1 tbsp cider vinegar (malt will do)
300g granulated white sugar

Strip the flowers from their stalks. Stir all ingredients together in a large, food-grade container, cover, and leave for 24 hours. Strain, simmer for 15 minutes, leave to cool and then pour into bottles. Leave for 3 weeks in a warmish place to allow time for the 'fizz' to develop. It's probably best to use small bottles with screw caps rather than corks – otherwise if there's too much fermentation the corks might blow off.

You can increase the volume, but keep the same proportions of ingredients.

Elderflower Cheesecake

This cheesecake is a based on a medieval recipe for a cream cheese tart called Sambocade.

SERVES 6–8

670g cottage cheese (or ½ cottage cheese
and ½ mascarpone)
2 tbsp dried elderflowers (simply hang
the flower heads, enclosed in large
paper bags, in a warm dry place for a
few days)
1 tbsp rose water (see page 166) or
orange blossom water
70g granulated white sugar
3 egg whites
1 x 20cm sweet pastry case, shop-bought
or home-made

Preheat oven to 190°C/gas mark 5.

In a bowl, mix together the cheese, elderflowers, rose or orange blossom water and sugar.

In a separate bowl, whisk the egg whites to soft peaks and carefully fold into the cheese mixture with a metal spoon. Spoon into the pastry case and bake for about 1 hour. When cooked, the surface should be slightly wobbly – the cake will continue to set as it cools.

Leave to cool before serving.

Elderflower Cake

MAKES 8 DECENT-SIZED SLICES

5 handfuls elderflowers
200g butter
200g honey
3 eggs, separated
400g plain flour
15g baking powder
¾ tsp ground cinnamon
100g almonds, finely chopped
50ml milk

Preheat oven to 180°C/gas mark 4. Grease an 8 inch cake tin.

Strip the flowers from their stalks. Beat the butter in a bowl until pale and creamy and add the honey, stirring well. Add the egg yolks and stir again until the mixture is smooth.

In a separate bowl, mix together the flour, baking powder, cinnamon and chopped almonds. Add to the butter mixture with the milk and mix together, then stir in the elderflowers.

Whisk the egg whites to soft peak stage and fold carefully into the mixture. Pour into the prepared tin, and bake for about half an hour.

Elderflower Sorbet

If you don't try any other recipe in this book, *please* make this sorbet. The world will instantly become a better place, any problems instantly fall away and you will be suffused with joy.

It's easier to use an ice-cream maker to make the sorbet, if you have one, but you can also make by hand.

700g caster sugar
1 litre water
8 elderflower heads
Zest from 3 lemons

For the sugar syrup, place the sugar and water in a large, heavy-bottomed saucepan over a moderately high heat and stir with a wooden spoon until the sugar has dissolved. Increase the heat and bring the syrup to the boil, then reduce the heat slightly so that the syrup is at a steady simmer. Allow the syrup to simmer for 5 minutes until it has thickened.

Remove the sugar syrup from the heat and add the elderflowers and lemon zest. Leave to infuse until the syrup has cooled.

Strain the syrup, pour into an ice-cream maker and prepare according to the manufacturer's instructions. Alternatively, pour into a container and place in the freezer for 2–3 hours. You will need to stir it every half hour or so, bringing the ice crystals from the outside into the middle, to stop the sorbet from crystallising too much.

Remove the sorbet from the freezer 10 minutes before serving.

Elderflower and Rhubarb Ripple

This recipe was devised specially for this book by Lucia Stuart, author of the lovely 'Eating Flowers' (www.eatingflowers.com)

Lucia has used rhubarb to make the ripple effect, but you could easily use blackberries, raspberries or wimberries instead.

RIPPLE
6 rhubarb sticks, cleaned and chopped
5 tablespoons of granulated white sugar

Put the rhubarb and sugar into a thick-bottomed pan with a lid. Place on a slow heat to dissolve the sugar, then simmer for 15 minutes. Allow to cool.

ICE CREAM
12 sprigs of elderflower blossom,
  stalks removed
300ml double cream
400ml full fat milk
100g caster sugar
6 medium egg yolks

Mix together the milk and cream. Shake the flowers clean, strip them from their stalks and put them to soak in the liquid overnight.

Next day, strain the mixture into a heavy-bottomed saucepan with the sugar.

Put over a medium heat, dissolve the sugar and bring to the boil. Take the pan off the heat, allow to cool for 10 minutes and then add the egg yolks, stirring vigorously to prevent lumps.

Return to the heat and cook slowly, stirring and whipping until the cream turns into a custard thick enough to coat the back of a wooden spoon. Strain the liquid into a plastic tub and put into the freezer.

After 2–3 hours give the ice cream a good whipping, either by hand or with an electric whisk. This will break up any ice crystals. Pop back into the freezer for another couple of hours.

Before the mixture is almost frozen, cut 2 troughs in the ice cream and pour the rhubarb into it. Don't stir! Pop back into the freezer, and when it's all completely frozen, scoop out in balls.

Elderflower Sabayon

See recipe for meadowsweet sabayon on page 126, but use elderflower cordial instead of meadowsweet juice.

ELDERBERRIES

What an amazing tree is the elder. As soon as you've finished making gorgeous things with its flowers, it's generous enough to start producing gorgeous, lustrous purple-black berries with colourful red stems.

You can do a lot with these berries, which contain lashings of vitamins C and A. The only note of caution is to ensure that you only use ripe berries; unripe ones can make you ill.

Perhaps the simplest recipe is the elderberry vinegar below; it's not only delicious but it has the added benefit of being very good for you, since it's full of vitamin C which is good for fighting colds and flu.

MAKES APPROX. 1 LITRE

Simply take 500ml white vinegar, add 350g elderberries (stripped from their stalks) and leave for 5 days, shaking or stirring occasionally. Strain away the liquid and add 350g white sugar for every 260ml liquid. Boil together for 10 minutes, then leave to cool before pouring into bottles.

You can use this vinegar as a good replacement for balsamic vinegar – and at a smidgeon of the cost.

Elderberry and Almond Pie

This pie is sensational served with a sharp crème fraîche, yoghurt or ice cream – or, if you're very organised, elderflower sorbet.

You'll need a 20cm pie dish or flan tin.

SERVES 8

250g frozen shortcrust or puff pastry, defrosted
20 decent-sized sprigs of elderberries
2 tbsp water
Up to 200g sugar (see note below)
200g ground almonds

Preheat oven to 180°C/gas mark 4. Grease your pie dish or flan tin. Roll the pastry out into 2 rounds – one that's large enough to form the base of the pie, and the other for the lid.

Make the syrup by washing the berries well and stripping them from their stalks. Place in a pan with the water and half of the sugar. Simmer for 10 minutes and add more sugar if needed. After 10–15 minutes, the syrup will have thickened, at which point let it cool and then strain through a fine sieve to get rid of any pips.

Line the bottom of the greased pie dish with the pastry base. Add the ground almonds, covering the base, then pour the syrup on top.

Now carefully press the pastry 'lid' on top of the pie, wetting both pieces of pastry around the edges so that they stick to each other. Make a small hole in the top of the pie (unless you have one of those piebirds that allows the steam to escape during cooking). Brush with water and sprinkle with more sugar. Bake for about 40 minutes, or until the pastry is crisp and golden.

Elderberry Flu Remedy

Did you know that the humble elderberry contains an incredibly effective antidote to more strains of flu than Tamiflu can cope with? That's right; a simple syrup made from the berries of the black elder tree, swigged once a day, could not only immunise you against flu, but tastes delicious too. The efficacious substance is called Sambucol, from the Latin name for the plant, *Sambucus nigra.*

The efficacy of elderberries as a treatment for flu is a relatively recent discovery. Extensive studies into Sambucol took place recently in Norway and, more significantly, in Israel, where the symptoms of two types of flu were cured in two days, whereas Tamiflu took four to five days (not to mention its nasty after-effects). Unsurprisingly, lots of pharmaceutical companies are jumping on the bandwagon and you'll be seeing more and more expensive elderberry-based remedies on the market.

Here's how to sidestep those corporate pill-pushers and make your own delicious flu remedy. You'll need to think ahead since elderberries are at their best prior to the main flu season.

Pick a couple of kilos of elderberries on a dry, sunny day. Wash them, drain them and strip them from their stalks.

Put in a pan and just cover with water. Bring to the boil and simmer until they are soft. This will take about half an hour.

Strain through a sieve; the seeds are harmless but have a bitter taste.

For every 600ml of liquid, add 450g sugar, the juice of 1 lemon and 10 cloves.

Return to the heat, add a 2cm piece of fresh ginger, and simmer until the sugar has dissolved. Boil hard for 10 minutes. Let the liquid cool, then fish out the ginger and cloves.

Store in the freezer, either in plastic bottles or freezer bags, or freeze in ice cube trays.

This stuff not only does you good, but tastes amazing. You could dilute a couple of the ice cubes in hot water to make a lovely hot toddy, or pour the syrup over ice cream. The anti-inflammatory and antioxidant properties will not only help prevent flu, but will soothe any symptoms that you might already have.

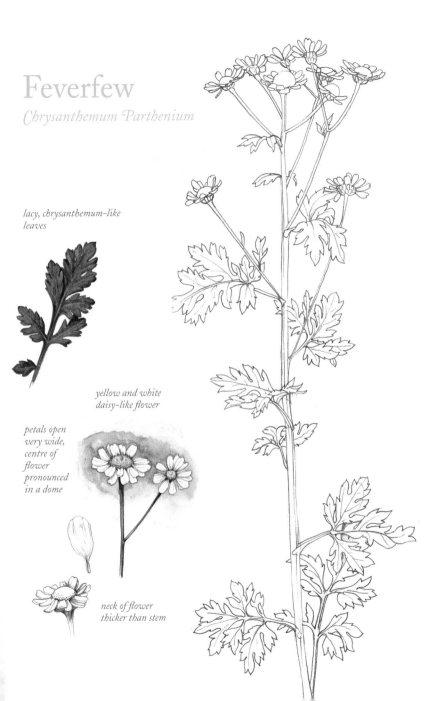

Feverfew
Chrysanthemum Parthenium

lacy, chrysanthemum-like leaves

yellow and white daisy-like flower

petals open very wide, centre of flower pronounced in a dome

neck of flower thicker than stem

Feverfew has yellow and white daisy-like flowers that grow in a shrub-like way, with lacy, chrysanthemum-like leaves, about 50cm high. A good way to identify feverfew is by its leaves, which have a pleasant smell that's a bit like camphor. The plant is perennial and quite invasive.

Culinary uses

Feverfew is known more for its medicinal properties than for its use in the kitchen. However, I found an unusual recipe from Italy. Feverfew cake has a delicately aromatic flavour and it's unlikely that your friends will be able to guess the ingredient that gives this cake its distinctive taste.

Medicinal uses

The Latin name for feverfew is *febrifugia*, which means 'fever reducer' and, as you might suppose, the plant helps reduce a high temperature and the sweats. Feverfew is known as an effective treatment against headaches, and in particular, migraines. In fact, research has shown that active ingredients in the plant stop blood vessels in the brain from going into the spasms that seem to be primary cause of migraines. If you're a sufferer, it's worth chewing on a few leaves every day and see what happens.

Feverfew is also effective against rheumatism and 'growing pains'. An infusion of feverfew not only acts as a tonic, having a pick-me-up effect, but can also help regulate the menstrual cycle.

Did you know?

One of the folk names for feverfew is 'bachelor's button'. The dried flower buds can be used as a general insecticide, and in particular, a moth repellent. So, if you have a problem with moths, a dried bundle of feverfew popped into your wardrobe will ward them off.

Feverfew Cake

You will need a 22cm cake tin – a spring form tin is ideal, if you have one.

225g plain flour
1 tsp baking powder
¼ tsp salt
110g sugar
1 tsp grated lemon zest
2 eggs
110g butter, at room temperature
284ml milk
110g finely chopped feverfew leaves

Preheat oven to 180°C/gas mark 4. Grease the cake tin.

Sift together the flour, baking powder, salt, sugar and zest. Pile the flour into a mound, make a well in the middle, add the eggs, one at a time, and mix with a fork. Add the butter, chopping it into the mix, and using a wooden spoon, stir in the milk. Add the feverfew leaves and stir thoroughly. Spoon into the tin and smooth the batter evenly.

Bake for about 40 minutes. Let the cake cool before removing it from the tin, and serve with fresh fruit and cream.

Ground Elder

Aegopodium Podagraria

pale yellow anthers

*leaves distinctly
asymmetrical*

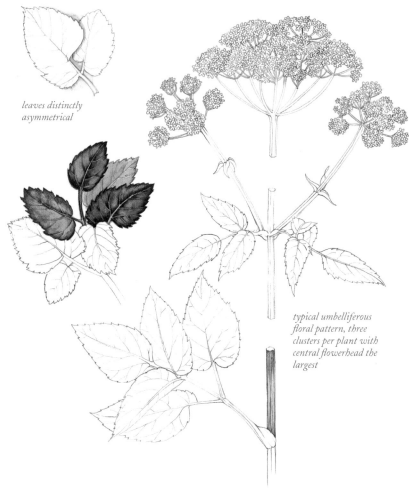

*typical umbelliferous
floral pattern, three
clusters per plant with
central flowerhead the
largest*

Ground elder grows close to the earth via creeping roots. The stems are hollow and the leaves divided into three leaflets. Pretty umbels of white flowers elevate themselves from those creeping roots in the summer months.

This low-growing, leafy plant has become the scourge of gardeners everywhere who try everything to eradicate ground elder from their gardens. It's incredibly invasive and notoriously difficult to remove.

Many keen gardeners might say that you're a very lucky person if you don't know what ground elder looks like. I would beg to differ, since it's a very tasty vegetable.

Culinary uses

There's a good reason why the Romans introduced the plant to Britain; the leaves and stalks are very tasty, slightly lemony and fragrant. In medieval times, the plant was actively cultivated as a vegetable. Cooking brings out a further peppery flavour. If you want to try the leaves on their own, go for young growth in early spring, before the plant has flowered.

Wash and cook the leaves in a little butter and a splash of water. Add salt and pepper to taste and you have a very good accompaniment to a meal. Ground elder is delicious chopped and added to potatoes before mashing them. And isn't it more constructive to use ground elder in this way rather than blasting it with toxic weed killers?

Medicinal uses

The *podagraria* part of the Latin name actually means 'gout', and the fresh leaves and dried root can be boiled into a concoction and used as a poultice to alleviate the painful symptoms. Otherwise, the plant has both diuretic and sedative effects.

Did you know?

Aegopodium means 'goat's foot', which is also one of the folk names for the plant. Other folk names include 'housemaid's knee', probably so-called because of the many hours the poor maid would spend on her knees trying to dig up the ground elder!

Ground Elder Pancakes

PANCAKES
110g plain flour
Pinch of salt
2 eggs
275 ml milk
Oil, and a knob of butter, for frying

FILLING
750g spinach
2 handfuls ground elder
Wild garlic leaves
Knob of butter
100g sour cream
150g goat's cheese
300g mozzarella

To make the pancake batter, whisk the flour, milk, eggs and butter together and leave to stand at room temperature for about a half an hour. Heat the oil in a frying pan until it's almost smoking, then pour off the surplus. Using a ladleful of batter per pancake, cook the pancakes one at a time until golden, flipping them over to cook the other side. Cover with a plate and keep them warm in a low oven.

To make the filling, tear up the spinach, the elder leaves and the garlic leaves. Melt the butter in a frying pan and cook the leaves until wilted. Stir in the cream and goat's cheese and cook for another 5 minutes, allowing the cream to reduce.

Turn up the oven to 180°C/gas mark 4.

In an ovenproof dish, stack the pancakes, adding a layer of filling in between each one. Top with slices of mozzarella and bake in the oven for about 15 minutes until the top is crisp and bubbling.

Alternatively, you can make individual filled pancakes, simply folded over with the mozzarella grilled on top. Delicious!

Ground Elder Soup

SERVES 4

10g butter
1 small onion, diced
1 dsp plain flour
500ml vegetable stock
2 bunches ground elder
250ml single cream
Salt and pepper, to taste

Melt the butter in a saucepan and soften the onion until golden. Add the flour and cook for a few minutes, stirring all the time. Add the stock a little at a time, stirring to make sure no lumps form. Add the elder leaves and simmer for 5 minutes, or until the leaves have softened. Pour into a blender and blend until smooth, then stir in the cream and season to taste. Serve with croutons or, perhaps, with some crusty home-made wild garlic bread.

Hawthorn

Crataegus Monogyna

frothy white blossoms
with heady scent

shiny bright red berries, or
haws, appear in autumn

leaf colour varies from
bright to dark dusty green

Also known as the maythorn, this is one of Britain's ancient trees, often used in hedging, but also found as a free-standing tree. The hawthorn – depending on its location – can grow up to 10 metres tall. A deciduous tree, its shiny lobed leaves start to unfurl in early spring. Its frothy white blossoms have a distinctive, heady scent; the tree was at one time usually blossoming by May Day, hence the name, but the calendar was revised in 1752, bringing the first of May forward by 13 days, so nowadays it's less likely to be in flower at that exact time. After the blossoms fade the haws start to develop, growing into shiny bright red berries by autumn, just when the tree starts to shed its leaves.

Incidentally, we know that the oldest hedges contain an earlier form of hawthorn, *Crataegus laevigata* (English or midland hawthorn) rather than the one we're dealing with here. The best way of identifying the English hawthorn is by its scent; whereas the *monogyna* smells heavenly, the odour of the English hawthorn has been described as 'putrid'.

The hawthorn is a fairy tree par excellence. When it grows together with the ash and the oak in close proximity, that place is meant to be steeped in fairy lore – a part of fairy-land itself. Because of its thorns, the tree is considered to be protective, and this includes against fire.

In days gone by, a globe of woven hawthorn would be brought into the home to ask the fairies to make sure that the house wouldn't be burned down. Every year the globe would be replaced and the old one burned (presumably, outside!).

The magic of this tree is further marked in the shape of its flowers. The five petals are considered to make a pentagram, itself a magical sign sometimes known as the Elven Cross. Like the elder, the heady fragrance of mayflowers, if inhaled deeply, is believed to help one access the 'other world'; and of course we've seen how the tree was named for the time its flowers blossom, on May Day, which is when the pre-Christian celebration called Beltane, traditionally dedicated to the fairies, took place. As the sap rises in the new trees and plants, so it does in human beings, and traditionally Beltane is a time of sexual congress. Could this mean that recipes using the leaves, fruit and flowers of the hawthorn might have aphrodisiac qualities?

Culinary uses

Your grandparents might remember eating the young leaves of the hawthorn. They're so filling that they used to be called 'bread and cheese' and have the added advantage of lowering cholesterol. These leaves can be added to salad or used as greenery in sandwiches, but the young fresh leaves are the tastiest. They're especially nice added to a salad of new potatoes and spring onions, tossed in a vinaigrette dressing.

Medicinal uses

Health practitioners of both Western and Eastern disciplines use the hawthorn in remedies for hypertension. The plant has compounds that help increase blood flow through the arteries, which helps alleviate the symptoms of arteriosclerosis and angina. The berries are also rich in vitamin C, and it's worth making the syrup recipe here to take to ward off colds.

Did you know?

Legend has it that the Glastonbury Thorn, the mysterious tree that flowers at Christmas and Easter, is descended from the staff that Joseph of Arimathea plunged into the ground, which then took root. The thorn was vandalised in December 2010; however, this isn't the first time that the sacred tree has suffered. Pilgrims and tourists have been taking cuttings for centuries, and the Puritans hacked it back severely to stop what they considered to be the blasphemous idolisation of a tree. However, all this 'pruning' has served not only to publicise the tree, but to make it stronger physically, too. Very apt, since *Crataegus* means 'strong'.

Mayblossom Sorbet

You need to ensure that the flowers are picked on a hot day and that they're young – the delicious light aniseed scent isn't quite so nice once it's aged slightly. Use the flowers straight away. Don't wash them, but do remember to shake them to remove any insects.

An ice-cream maker is useful, but not essential, for this recipe.

SERVES 4

700g caster sugar
1 litre water
20 mayblossom clusters
Zest of 3 limes

To make the sugar syrup, place the sugar and water in a large saucepan over a moderately high heat and stir with a wooden spoon until the sugar has dissolved. Increase the heat and bring the syrup to the boil. Reduce the heat slightly so that the syrup is at a steady simmer. Allow the syrup to simmer and thicken for 5 minutes. Remove from the heat and add the flowers and lime zest. Leave to infuse until the syrup has cooled.

To make the sorbet, strain the syrup and either pour into an ice-cream maker and follow the manufacturer's instructions, or pour into a freezer container and freeze for 2–3 hours, stirring every half hour or so to stop the sorbet from granulating too much.

Remove the sorbet from the freezer 10 minutes before serving. Serve with a garnish of leaves and blossoms.

Hawberry Brandy

Hawthorn berries (haws), too, can be used in a variety of recipes. In fact, they're so prized in India that the trees are cultivated specifically for their fruit. Here's one of the easiest recipes.

MAKES 1 X 70CL BOTTLE

70cl bottle of your favourite brandy
450g haws, washed well and dried
 (they can be surprisingly dusty)
Sugar, to taste (optional)

Divide the brandy between 2 empty brandy bottles. Add sugar, if you're using it, and then the haws, dividing them between the 2 bottles. Leave enough room at the top of each bottle to allow you to shake the contents from time to time. Leave for 3 months, shaking the bottles occasionally, then remove the fruits and reunite the brandy in one 70cl bottle.

Haw Syrup

MAKES APPROX. 4 LITRES

1kg haws, washed
3 litres boiling water
450g granulated white sugar

Mash up the haws with a wooden or stone implement (a metal one would reduce the vitamin C content and this, after all, is what you're after). Put the haws in a large (earthenware if you have one), flameproof pot and add the boiling water. Simmer for 20 minutes, strain the liquid from the haws into an earthenware or glass bowl. Discard the haws since there's not much else you can do with them at this point, then pour the syrup back into the pot, add the sugar and boil again for 10 minutes or until the liquid has reduced and thickened to a syrupy texture.

Pour into warmed, sterilised jars or bottles, or alternatively let the syrup go cold and decant straight into cold containers.

If you have a cold, dilute 1 part syrup with 1 part hot water, add a squeeze of lemon and a tot of brandy. Drink before bedtime and, with any luck, you'll wake feeling much better.

Hawthorn Fruit Leather

This recipe is from Sarah Howcroft, shaman and bushcraft trainer. A dehydrator, if you have one, is useful for drying the fruit purée here. Many domestic ovens can also be used for dehydrating, using a cool setting, but, be aware that the purée will take about 12 hours to dehydrate and the oven will not be available for anything else during that time.

SERVES 10 AS SWEETS OR NIBBLES

Large quantity of washed hawthorn berries (the larger and more tender the better)
Boiling water, for scalding
Small cup of boiling water, for mixing

Scald the hawthorn berries first by placing them in a sieve and pouring boiling water over them. Transfer the berries to a bowl and, using a potato masher or similar, mash them with a small amount of hot water. You want to achieve a pulp that is not too wet, but able to be separated from the stones. Purée the mixture by using a large spoon to push the pulp through a sieve, leaving the hard seeds behind.

Spread the purée to about 3mm thick, on dehydrator sheets or greaseproof paper, and, if you happen to have one, dry it thoroughly in a dehydrator, or in the oven on a cool setting, for 12 hours. It is vitally important to dry the mixture completely or it will go mouldy later. Then you can cut it into strips and roll them up, or into squares.

You could also mix other fruit purées, such as apple, with the hawthorn, for different flavours.

Hazel
Corylus Avellana

young leaves shiny, crinkled and
flushed with crimson which fades
towards the edges and with age

nuts start pale green, flushed
brown at base, becoming
brown all over with age

An important tree of Britain's native woodlands, the hazel rarely grows beyond 6 metres and can often be found tucked away in hedgerows. The leaves are circular in shape, but with jaggedy edges, pale green in colour. The catkins, called 'lambs' tails', appear in the spring. These catkins are in fact the male flower. The female flower is no more than the size of a small bud, but it's the male catkin that's most visible, dangling to a length of 8cm. The nuts come in bundles of three, and are a pale cream when they first appear.

The old Irish Gaelic name for the hazel is 'coll', and so sacred was the tree that one of the three Celtic kings of Ireland, McCuill (meaning 'son of the hazel') was named after it. The tree stands where the 'real' world and the 'other world' join, and a tree spirit is said to protect the tree – especially its nuts. This spirit has different names in different parts of the UK, and in Yorkshire she was named Churn Milk Peg. In fact, every nut tree had its supernatural guardian or protector; the reason being that nuts are not only symbols of fertility, but are also full of goodness and nourishment and, importantly, they're easy to store against the lean winter months.

Culinary uses

If you're harvesting hazelnuts, you'll be lucky to beat the squirrels to them! It's pointless collecting them before they are properly ripe since they don't ripen very well off the tree. Unripe nuts range from pale cream to green in colour, whereas ripe ones are brown. Cracking the nuts is easy; the shells aren't too hard. Use a normal nut cracker. To avoid 'crushed nuts' put the nut lengthways, not *horizontally,* in the nut cracker so you apply pressure from the tip and base (not the sides) and you should get a complete kernel every time (thanks to Lynne Allbutt for this tip). Otherwise, put the nut on a hard surface and simply rap on the side with a heavy stone or similar implement.

If after harvesting the nuts and eating a few you still have some left, here are some ideas for what to do with them.

Medicinal uses

Powdered hazelnuts, mixed with honey and water, can soothe a troublesome cough. Those nuts are also a good source of calcium, protein and potassium. A tea made of hazel leaves can help regulate menstrual flow, and is particularly helpful in alleviating heavy periods.

Did you know?

Forked hazel twigs shaped like a letter Y are favoured by dowsers, who have the talent to find water and other substances that are hidden from view by holding the twigs and responding to the movements they make. How and when dowsing started is subject to debate, but the ancient Egyptians and Babylonians are believed to have practised it. It's possible that hazel is a popular choice for use as a dowsing rod because of its existing magical reputation.

Spicy Caramelised Hazelnuts

225g hazelnuts, shelled
Knob of butter
1 chilli pepper, de-seeded and chopped
110g granulated white sugar
1 tbsp soy sauce

Keeping the nuts as intact as possible, toast them under a grill, shaking them occasionally.

Heat a heavy-bottomed frying pan, add the butter and sauté the chilli. Add the sugar, stirring to keep it from burning, and add a dash of soy sauce as the sugar starts to melt. Once the sugar starts to brown, toss in the toasted hazelnuts and keep stirring. Add more soy sauce as needed. Once the sugar has caramelised and the hazelnuts are nicely coated, take the pan off the heat and allow to cool.

These are delicious served with drinks.

Italian Hazelnut Biscuits

MAKES ABOUT 20 BISCUITS

200g butter, softened
200g granulated white sugar
2 medium eggs, beaten
1½ tsp vanilla extract
230g plain flour
1 tsp ground allspice
½ tsp salt
¾ tsp baking powder
175g (6oz) chopped hazelnuts (or you
  can use half hazelnuts and half
  chocolate chips)

Preheat oven to 180°C/gas mark 4
and grease a large baking tray.
In a warmed bowl, cream together
the butter and sugar. Add the beaten
eggs and vanilla, and whisk. Sift
together the flour, spice, salt and
baking powder and stir into the
mixture until well combined, then
stir in the chopped hazelnuts.

On a lightly floured surface, roll
the dough into two tubular lengths
of 30cm, then flatten to about 1cm
thickness. Transfer to the prepared
baking tray and bake for about half
an hour. Take out of the oven, cut
into 2cm-wide slices and bake for
another 10 minutes. You'll need to
test the biscuits at this point to make
sure they're hard and crunchy.

These are fabulous dipped in coffee
or with thick hot chocolate.

Fallen Hazelnut and Cheddar Crackers

This recipe, which uses fallen hazel-
nuts, comes courtesy of Allie
Thomas, proprietrix of Cradoc's
Savoury Biscuits in Brecon.

Like fallen angels, these hazelnuts
seem to have slightly more personal-
ity and attitude than the ones that
stay up in the clouds, so to speak,
feeling all superior. But you'll have to
keep a sharp eye out for the squirrels
who will know the precise moment,
to the second, that the nuts are ready
to harvest! This is a crisp, melt-in-
the-mouth biscuit, with a nutty bite,
that's particularly good with blue
cheeses and soft white curd cheeses.

You'll need to collect enough nuts
to make 100g of kernel, so this will
be about twice that weight in un-
cracked nuts (plus a few over for
popping into your mouth). Next,
pulse them in a food processor for a
few seconds; don't forget to leave a
few interesting lumpy bits.

MAKES APPROX 20 BISCUITS

65g butter
250g best-quality self-raising flour
100g strong cheddar (the stronger the
  better), grated
100g chopped hazelnuts
¼ tsp mustard powder
¼ tsp white pepper (optional)

Good pinch of paprika
Good pinch of celery salt
1 egg, lightly beaten
Milk, to bind (optional)

Using your fingers, rub the butter and flour together to form powdery crumbs. Add the grated cheese, chopped nuts, mustard powder, white pepper, paprika and celery salt. Mix well with a metal spoon, then add the beaten egg to make a firm dough, adding a little extra milk if you need it.

Wrap the dough in clingfilm and chill in the fridge for 20 minutes. Meanwhile, preheat oven to 180°C/ gas mark 4 and grease a large baking tray.

Remove the dough from the fridge and divide into four. Flour a cool work surface and roll out the first quarter of dough, squashing the nuts gently as you roll (this gets easier after the first time). Roll out the dough as thinly and evenly as you can. The cooler the dough, the easier this is. Repeat with the remaining dough.

Either cut into strips or squares or use a pastry cutter to cut into rounds. Prick with a fork. The biscuits will not expand during cooking, so lay them edge to edge on the prepared baking tray.

Bake for 15–18 minutes, until the biscuits are dry and crisp and browned on the edges. Cool on a wire rack. The biscuits will keep for up to 5 days in an airtight tin.

Himalayan Balsam
Impatiens Glandulifera

young seed pod

*blooms vary from pink/white
to dark purple*

black seeds

*leaves bright glossy
green, red serrated
edges, parallel
alternate veins*

ripe seed pod

exploded seed pod

Quite a controversial plant, this. Proliferating along riverbanks and other shady places prone to damp soil, the plant was introduced to the UK in 1839 and is now considered to be 'naturalised'. But is it, strictly speaking, a traditional hedgerow plant? No, but the recipes here use the seeds and, by harvesting the seeds, you might help stop the plant spreading, even in a relatively small way. Himalayan balsam spreads rapidly, is quick to take over from native plants, and so is regarded as a pest.

Despite this, you can't deny its beauty. It can grow up to nearly 3.3 metres high and has pretty pink flowers, a bit like snapdragons. In late summer, the seed pods of Himalayan balsam burst at the slightest provocation, scattering seeds far and wide in a joyous and surprising explosion. And here's the problem. One plant can produce somewhere in the region of 2,500 seeds. There are various methods of eradicating the plant, but since it's here, we might as well find ways of using it.

Culinary uses

The leaves of the plant are peppery and taste good in salads. The stems, too, are edible and can be steamed and served with butter and lemon juice, like asparagus.

If you like, you can simply pick the seed pods and nibble on them, whether they are ripe or not.

Harvesting the seeds: The last thing you want is to disperse seeds as you harvest them, so take them just before the popping season starts in earnest – this would be in early autumn. If you miss this window of opportunity, then take a large paper bag and place it over the top of the stem. Pinch the bag together at the opening, then tap the bag to knock the seeds off the plant and into the bag. The raining of the seeds on the inside of the bag is highly satisfying and a lot of fun for children!

If you are harvesting the seeds without bursting the pods, try to remove as much of the green fibre of the actual pod as you can.

The seeds can be ground and roasted to make a drink, or dried and ground to make a sort of gluten-free flour.

Medicinal uses

Also known as 'jewelweed', the balsam is used as a Bach flower remedy to cure mental tension, irritability, and impatience.

Did you know?

The folk names for Himalayan balsam include policeman's helmet, bee bums, jumping jacks, and poor man's orchid.

Himalayan Balsam Seed Curry

This is my favourite recipe for using the balsam seeds. It not only tastes amazing but helps prevent the spread of the plant. It was devised by Robin Harford of www.eatweeds.co.uk and I'm very grateful that he let me include it here.

SERVES 4

Good glug of olive oil or ghee
1 large onion, chopped
2 garlic cloves, crushed
2 chilli peppers, de-seeded and chopped
Chunk of fresh ginger, grated
Approx. 4 tbsp Himalayan balsam seeds
   (collect as many as you can; see above
   note on harvesting)
2 sticks celery
1 large swede
1 red pepper, de-seeded
1 yellow pepper, de-seeded
2 tbsp curry paste
2 large tomatoes
1 400g can chopped tomatoes
1 sachet coconut milk, made up
   according to manufacturer's
   instructions

Melt the ghee in a large frying pan. Add the chopped onion, crushed garlic and chilli and sauté for a few minutes until the onion is softened. Add the balsam seeds and stir until the seeds are coated.

Chop the rest of the vegetables (including the tomatoes) into about 2cm cubes and add to the pan. Add the ginger too and stir for 5 minutes, then add the curry paste and the tinned tomatoes and simmer to reduce. Stir in the creamed coconut, then reduce the heat and cook for a further 5 minutes. Allow to stand for another 5 minutes before serving with rice and/or naan bread.

Himalayan Balsam Seed Rissoles

SERVES 4

1 small onion, chopped
125g Himalayan balsam seeds
50g breadcrumbs, fresh
50g strong cheddar cheese
1 tbsp sundried tomato paste
1 egg
Good pinch of dried herbs
Dash of soy sauce
Salt and pepper, to taste
Butter and olive oil, for frying

Preheat oven to 200°C/gas mark 6. Smear an ovenproof dish with a little olive oil.

Heat a mixture of olive oil and butter in a frying pan. Fry the onion, stirring to prevent it from scorching and, when the onion is translucent, throw in the seeds. Cook over a medium heat for a couple more minutes.

Leave to cool a little, then transfer the onion-seed mixture to a large bowl and add all the other ingredients. Bring the mixture together with your hands and shape into short, fat sausage shapes. There should be enough to make about 12. Transfer to the ovenproof dish and bake for about half an hour, turning the rissoles occasionally.

Serve with a thick garlic and tomato sauce: simply put 2 cans of chopped tomatoes and a couple of crushed garlic cloves in a pan over a medium heat and cook until reduced. Add salt and pepper to taste.

Honeysuckle
Lonicera Periclymenum

*one main stigma,
five stamen*

*yellow-pink trumpet-shaped
honeysuckle blossom*

The honeysuckle that's so popular in cultivated gardens has its smaller, wilder counterpart in the plant that can be found twirling its way upwards through the trees and shrubs of hedgerows and verges. You often smell honeysuckle before you see it since the flowers tend to be hidden away amongst the foliage of other plants. The scent of honeysuckle, though, is unmistakably divine. The yellowy pink, trumpet-shaped flowers are followed by clusters of bright red berries.

Culinary uses

Honeysuckle flowers are not only edible but also quite delicious, and look gorgeous thrown into salads or used to decorate desserts. The berries, however, are poisonous, so leave them well alone. One of the old names for the honeysuckle is the delightful 'woodbine', because of its habit of twining itself around trees.

The recipes here mainly use honeysuckle flowers, although the leaves are also used in the recipe for honeysuckle tea below, which, as well as being an expectorant, is a delicious drink in its own right.

Medicinal uses

Honeysuckle contains salicylic acid, the main ingredient in aspirin. Honeysuckle-flower syrup can help ease coughs and sore throats and help you get rid of phlegm – the technical term is 'expectoration'. The plant contains natural antiseptics and can be used as a poultice for skin infections.

Did you know?

The gorgeous scent of honeysuckle is stronger in the evening than during the day – it is said that moths can smell it from a quarter of a mile away.

Honeysuckle Sorbet

You *must* try this – it's a real taste of summer. If possible, keep some in the freezer to enjoy in the depths of winter. Pick the blossoms on a hot day when the sun has been on the flowers for a few hours.

An ice-cream maker is useful for this recipe, if you have one.

SERVES 6

*850g fresh honeysuckle blossoms, tightly
  packed but not crushed
Just under 1.5 litres cool water
Small pinch of allspice*

SYRUP
*380g sugar
500ml water
1 tsp fresh lemon juice
Allspice or cinnamon powder*

Put the flowers into a large bowl, add the cool water, cover, and leave to stand overnight.

To make the syrup, put the sugar and the 500ml water into a heavy-bottomed pan and stir over a low heat until the sugar has dissolved. Bring to the boil, and boil for a couple of minutes to thicken into a syrup. Turn off the heat and add the lemon juice. Leave to cool.

Strain the water from the flowers into a jug, squeezing every last drop of juice out of them. Add the infusion to the syrup with a small pinch of the spice you prefer. Pour the mixture into an ice-cream maker, if you have one; otherwise freeze the mixture in a glass dish for an hour or so, then mash up with a fork, and return to the freezer. Repeat twice more. Next blend with a hand blender so that it looks like snow, then return the mixture to the freezer for about a half an hour or so. Take out and leave to stand for a couple of minutes before serving.

Garnish with honeysuckle blossom and mint leaves.

Honeysuckle Tea

SERVES 4–6

1 litre water
360g honeysuckle flowers and leaves,
  dust and debris rinsed off
Honey, to taste

Bring the water to the boil in a pan
and throw in the honeysuckle flowers
and leaves. Simmer for 10 minutes to
infuse the plant matter, then strain
the liquid into a jug. Put the water
back into the pan and add the honey.
You don't need to have a sore throat
to enjoy this drink!

Honeysuckle Jelly

This can be left to set in a shallow
tray, then cut into cubes and dusted
with a mix of icing sugar and cornflour, like Turkish delight.

A jelly bag is ideal for straining the
'juice', or you could use several layers
of cheesecloth.

1.8 litres honeysuckle 'juice', made with
  2kg fresh honeysuckle flowers
2 tsp lemon juice
2kg jam sugar with added pectin

To make the juice, lightly rinse the
flowers, cover with water in a large
pan and bring to the boil. Cover,
lower the heat and simmer for a further 20 minutes. Cool and refrigerate
overnight or for a couple of days if
you want a stronger honeysuckle
flavour.

Strain the liquid into a large bowl
through a jelly bag, if you have one,
or use several layers of cheesecloth
loosely draped over the top of the
bowl, fastened securely with elastic
bands. This isn't a fast process, so allow it to drip for a few hours, or even
overnight.

Transfer the strained liquid to a
large pan, add the lemon juice and
the sugar and bring to a rolling boil.
Boil for 1 minute, stirring constantly.
Then do a 'setting' test; put a blob of
the mixture on a chilled saucer, and

if it's ready then the blob should wrinkle up on the top when you push it with your finger. Remove from the heat and, once it's cooled a little, remove any scum from the surface with a slotted spoon. Pour the jelly into shallow trays to set. When set, cut into squares as described above. Alternatively, you can pour into warmed, sterilised jars, and eat with sharp cheeses or with ice cream.

Lady's Smock
Cardamine Pratensis

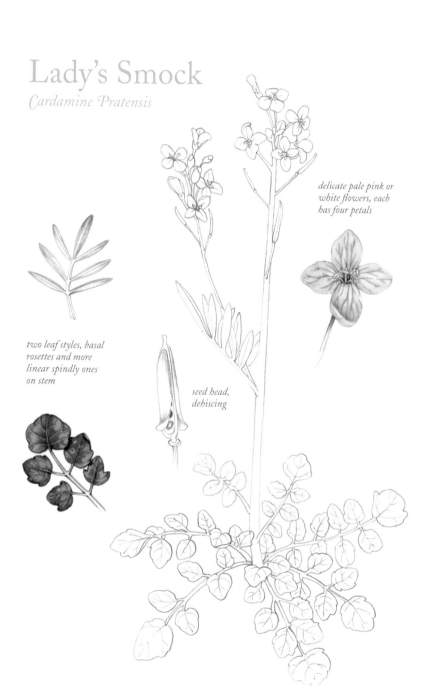

delicate pale pink or
white flowers, each
has four petals

two leaf styles, basal
rosettes and more
linear spindly ones
on stem

seed head,
dehiscing

Lady's smock is a very pretty, delicate meadow flower, also known as the cuckooflower because it blossoms at about the same time that the cuckoo is heard, in April and May. It can grow to about 60cm tall at its highest but is often smaller than that. It has very delicate pale pink or white flowers, each with four petals, and the plant boasts two types of leaves: ones that symmetrically climb the stems and are spindly like pine needles, while the leaves at the base are circular, again spaced symmetrically and graduated in size.

Culinary uses

Lady's smock is sometimes called bittercress, which is an accurate description of its appearance and flavour. It was once grown specifically as a salad herb, but it seems to have gone out of fashion as a cultivated plant, which would be a real loss were there not plenty of them available in the wild.

The leaves of the plant are very spicy and a little goes a long way, in terms of flavour. Try them added to new or mashed potatoes, thrown into salads, or use to add a special flavour to soups.

You might also try making a delicious herby butter: take a decent brand of unsalted butter, allow it to warm slightly and then mash up with a good pinch of finely chopped leaves. This is sensational on hot freshly made bread and has a very sophisticated flavour, a bit like horseradish. Indeed, the spicy leaves and flowers make a good substitute for it.

Medicinal uses

If you know a little Latin you'll recognise that the 'card' part of the name *Cardamine* refers to the heart. The plant was once believed to be beneficial to the heart. An infusion of the flowers and stems of lady's smock both stimulates the appetite and eases indigestion. Rich in vitamins and minerals, including potassium, iron and magnesium, lady's smock is a member of the mustard family. Mustard oils stimulate the circulation of the blood to the skin, the liver and the kidneys, and also relieve rheumatism.

Did you know?

There are many other flowers that have the folk name of cuckooflower, purely because of the time of year they come into bloom. These include the wood anemone and the bluebell, which is why I chose to use the name lady's smock for this entry. The name is said to be because of the smocked criss-cross effect of the veins in the tiny petals.

It wasn't so very long ago that people believed the lady's smock belonged to the fairies – another of its folk names is the fairy flower – and for this reason it was considered bad form to use it in any May Day revels, despite its abundance at that time of year. There was also a curious superstition that the plant attracted poisonous snakes, but this is probably because our native snakes share the same liking for grassy meadows as lady's smock. Are you a keen spell caster? If you want to make a love spell, you'll need to dig up the tubers of the lady's smock, since they are one of the prime ingredients for such magical practices. If you want to know what to do with the tubers once you have them, ask a witch!

Lady's Smock Leaf Salad

Mix together fresh young dandelion leaves, lady's smock, and watercress. Serve with a simple dressing of olive oil and lemon juice.

SERVES 2

Olive oil, for frying
30g fresh young nettle tops
30g lady's smock leaves, plus a few
   flowers for decoration
100g crème fraîche
120g strong Welsh goat's cheese, chopped
120g Caerphilly cheese, grated
Black pepper, to taste
2 large slices of homemade crusty bread

Heat the oil in a pan and add the
nettles and lady's smock leaves. Cook
for a few minutes until wilted. Cool,
then chop and mix with all of the
crème fraîche and half of each of the
other cheeses and some black pepper.
   Toast the bread very lightly on
both sides, then spread each slice
with the mixture, top with the rest of
the cheese, and grill until bubbling.
Serve with more black pepper and a
small sprinkling of the flowers.

Laver
Porphyra Umbilicalis

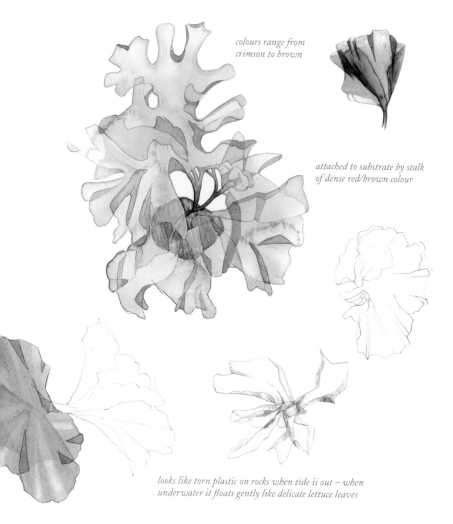

*colours range from
crimson to brown*

*attached to substrate by stalk
of dense red/brown colour*

*looks like torn plastic on rocks when tide is out – when
underwater it floats gently like delicate lettuce leaves*

Although this seaweed is synonymous with laver bread, one of the national dishes of Wales, it can be found in other parts of the UK, too. It appears from March onwards and looks almost like pieces of wet, crumpled plastic bags, coloured a browny-purple or even green, attached to the rocks by a small circular sucker.

Culinary uses

Laver is incredibly good for you; it's stuffed full of vitamins A, B, C and D, as well as being very low in calories. But both the taste and texture might be said to be an acquired taste!

When you harvest the laver try to collect as little sand as possible, and wash it very thoroughly. No matter how well you wash it there always seem to be a few teeny gritty bits that get between your teeth, but this is part of the beauty of an authentic, seaweed-gathering experience that you just can't fake!

Medicinal uses

Sailors used to eat laver on long sea voyages, to counteract scurvy, which can be caused by a vitamin C deficiency.

A poultice of laver, boiled in its own juice, has a remarkable effect on cuts and wounds.

Did you know?

You may be more familiar with nori seaweed, used in Japanese cuisine and which is sold as flat, dried sheets. This is in fact laver, which is the generic name for seaweeds of the *Porphyra* family.

The name for this traditional dish is a little misleading as it's not a bread at all, but rather a thick, sticky purée made from the seaweed. It is very simple to make.

Wash the seaweed thoroughly, tearing the larger 'leaves' into smaller pieces. Put into a large pan, cover with a little water, put the lid on the pan and then leave it to cook, very slowly, for a few hours until it's broken down into a smooth, soft, almost black purée.

Pour away any excess liquid and store in an airtight container in the fridge, where it will keep for about a week. This fresh laver bread knocks the socks off the tinned variety you can sometimes pick up in supermarkets.

An old way to serve laver is to spread it on top of a piece of hot bread and dripping, although this would completely cancel out the benefits of it being a low-calorie food! The following recipe is a different twist for a fabulous brunch-time snack.

MAKES 4 SLICES

Good knob of butter
8 large portobello mushrooms, peeled
  and cut into rough chunks
3 garlic cloves, crushed
½ tbsp sundried tomato paste
8 tbsp laver bread
2 tbsp single cream
10 spring onions, chopped
Salt and freshly ground pepper
4 large slices fresh crusty bread, to serve

Melt the knob of butter in a frying pan and sauté the mushrooms and garlic for 3 minutes. Stir in the sundried tomato paste and cook for another minute or so. Add the cream, reduce for a further couple of minutes. Take off the heat, stir in the laver, heat again for 1 minute and then throw in the spring onions right at the end.

Dollop onto the crusty bread, give it a cursory spread and season liberally with salt and pepper. Sensational!

If you have a few lady's smock flowers to hand, add them, too; their mustardy piquancy really adds to the dish.

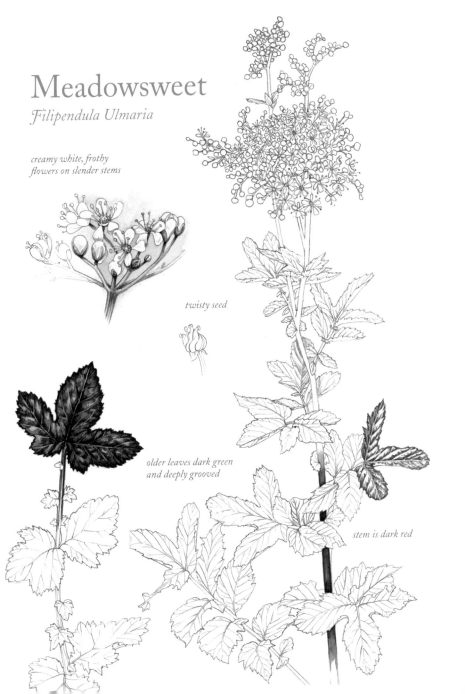

Meadowsweet

Filipendula Ulmaria

creamy white, frothy
flowers on slender stems

twisty seed

older leaves dark green
and deeply grooved

stem is dark red

The hybridised version of this glorious plant has infiltrated our gardens under the name of astilbe, which comes in a variety of colours. The wild plant, though, bears only creamy white, frothy flowers on top of slender stems that can grow up to 1.5 metres high. The leaves of the meadowsweet are oval and jagged, climbing in symmetrical pairs to the tops of their stems, where a lone leaf takes the top position. It's the scent of the flower that will probably first attract you to it, though; rather like a spicy vanilla, it's as fragrant and heady as elderflower and, indeed, meadowsweet can be cooked and prepared in all the same ways as elderflower.

Culinary uses

The blossoms can be substituted for elderflower in recipes. The leaves taste good, too, with a slightly aniseedy flavour, and make a great addition to soups and salads.

Medicinal uses

The leaves of the meadowsweet plant contain salicylic acid, the active ingredient in aspirin. Willow bark contains salicylic acid, too, but the chemical was first isolated from the meadowsweet plant. It's certainly worth drying some of the leaves to use in case of a headache, made into a tea. Just collect them on a warm,

dry summer's day and hang in bunches in a warm, dry place, bagged loosely in paper bags to catch any falling debris. Once they're dried, the leaves can be crumbled into dark jars to preserve them.

To make meadowsweet tea, infuse a couple of teaspoons of the crumbled leaves in a pot for 5 minutes and flavour with honey or sugar.

The plant also contains anti-rheumatic compounds. The flowers are used in an infusion to treat colds and flu, fluid retention and arthritis. The antiseptic qualities mean that an infusion is also good for relieving the symptoms of urinary tract infections, such as cystitis.

Did you know?

Meadowsweet is one of the sacred Druid plants and its scent is so stupefying that, according to an old superstition, it can lull you into a sleep from which you may never awake. Nevertheless, in Elizabethan times it made a very popular 'strewing' herb, scattered on the floors of houses to make them smell pleasant.

Meadowsweet Mead

MAKES APPROX. 3 X 70CL BOTTLES

Meadowsweet is so-named because it was once very popular in adding a subtle flavour to mead (in fact, the old English name for the plant was 'Meadsweet') and it's possible to do this very easily as long as you have access to a friendly beekeeper with a quantity of honey.

Take 1.4kg freshly gathered honey, wax scrapings and all; the 'scraps' from a honey-making session will do nicely. Put them into a large food-grade plastic bucket, then add nearly 7 litres water that's been boiled and allowed to cool to blood temperature, and 9 heads of meadowsweet flowers. Leave, covered, in a warm, dark place for about a week, then strain the liquid (which should have started fizzing as a result of the fermentation process) away from the solids, and filter into a demijohn. Seal with a wine-making airlock and then put the demijohn into a warm, dark place for another 3 months.

You will notice that a cloudy 'must' has formed at the bottom of the demijohn. Siphon the liquid into sterilised bottles, being careful not to disturb the sediment. You shouldn't need yeast, by the way, but if fermentation is slow you can add a sachet, following the manufacturer's instructions. They say that mead is best after 7 years, but who wants to wait that long?

Meadowsweet Sabayon

A sabayon is a delicious fresh egg custard that's usually made with eggs, sugar and wine. Here, the wine has been replaced with meadowsweet 'juice'.

PER PORTION
1 egg yolk
1 tbsp white caster sugar
1½ tbsp meadowsweet 'juice' (see instructions below)
Zest of 1 orange

To make the meadowsweet juice, take 10 heads of meadowsweet flowers, stripped from their stalks. Cover with 4.5 litres water and add 280g sugar. Bring to the boil, then remove from the heat and allow to stand for 4–5 hours. Strain the 'juice' from the flowers. This will make approximately 4.5 litres 'juice'.

   To make the sabayon, whisk the egg yolk in a metal bowl (which later will be heated over a pan of water, like a bain-marie). Add the sugar, the cooled meadowsweet juice and the orange zest. The idea is to get plenty of air into the egg yolks, which will then begin to froth.

   Place the bowl over a pan of simmering water or pour the mixture into a bain-marie and continue to whisk, taking care that the eggs don't start to actually cook. Once done, pour the custard over freshly picked summer fruits, add a garnish of meadowsweet and serve.

(Water) Mint

Mentha Aquatica

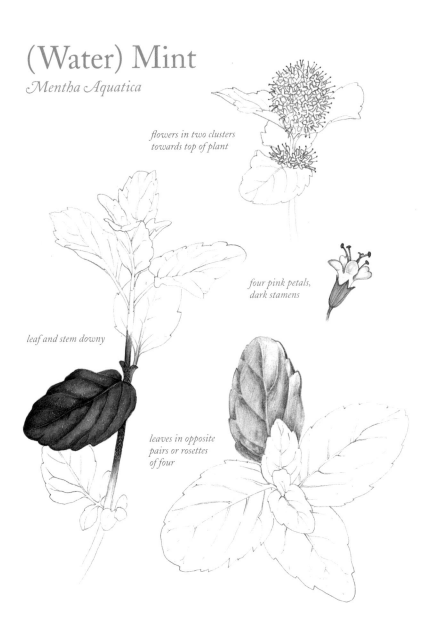

*flowers in two clusters
towards top of plant*

*four pink petals,
dark stamens*

leaf and stem downy

*leaves in opposite
pairs or rosettes
of four*

This type of mint grows, as the name suggests, in water and close to the edges of water. It can reach a height of 60cm. The leaves are oval, with blunt teeth along the edges. Long clusters of lilac blossoms form upright flower heads at the top of the plant. The growing season for water mint is quite long, ranging from June to September and even into October if the weather is mild.

Culinary uses

I've chosen water mint here because it grows in great profusion and, provided you leave the roots intact, you can pick a large quantity of it and really go to town cooking with it. There's nothing quite like going barefoot down to the stream where the water mint grows, your toes squishing through the mud, the scent of the crushed leaves drifting in the air, dragonflies leaping ahead of you.

Water mint freezes and dries beautifully, although the flavour is not quite as strong as regular garden mint. The flowers, too, are edible and can be used in salads, sprinkled on soups (traditionally, mint and pea go very nicely together), or frozen into ice cubes.

Be careful to wash water mint thoroughly, since its preferred habitat tends to be muddy.

Medicinal uses

As an infusion, water mint is very good for settling upset stomachs and for relieving the symptoms of colds and flu or painful menstruation.

Did you know?

In Elizabethan times, mint was used as a strewing herb and laid on the stone and wooden floors of houses. This would have had the added benefit of keeping creepy crawlies away – as the mint was crushed underfoot, its insecticidal oils would have been released.

Mint Tea

The easiest thing to do with mint is to make mint tea. It helps if you have some of those colourful, gilded traditional Moroccan glasses! Just add boiling water and sugar to taste to a few sprigs of water mint. Delicious made in large jugs, allowed to cool, strained, and served over ice with more fresh mint.

Minty Sugar

In this recipe the quantities of both ingredients will depend on the size of your jar.

A couple of big handfuls water mint leaves, washed well
Caster sugar

In a large, sealable jar, place alternate layers of caster sugar and mint, until it's all used up. Seal the jar and leave in a dark place, then after a month or so remove the mint leaves. You now have a delicious mint-infused sugar.

Mint Gulkand is similar to minty sugar (see the recipe for Gulkand on page 169 and make in the same way, substituting mint leaves for the rose petals).

Mint and Cucumber Lemonade

This recipe was inspired by a drink I had in Kerala, southern India, a couple of years ago. (Unfortunately the small café where I had it didn't have a name.) It took a few goes to get the proportions right but this is as close to it as I could get.

MAKES 1 LARGE JUGFUL

225g water mint leaves, plus 1 handful
500g caster sugar
250ml water
1 large cucumber
600ml sparkling water
Pinch of salt (optional)

Wash the mint leaves well. Melt the sugar in the water with the 225g mint leaves. Bring to the boil and simmer for 5 minutes, then strain the syrup into a jug and allow to cool completely.

Peel the cucumber, chop into chunks, put into a blender with the syrup and a further handful of mint leaves, and whizz until smooth. Pour into a large jug, add the sparkling water and pour over ice. A little salt adds a certain something, too.

Mint Chutney

MAKES APPROX. 6 x 454ML JARS

450g granulated white sugar
Dollop of grain mustard
450ml cider vinegar
450g eating apples, peeled, cored and
  finely chopped
2 medium onions, finely chopped
225g fresh water mint leaves
75g seedless raisins
Salt and pepper, to taste

Dissolve the sugar in a heavy-bottomed pan with the mustard and vinegar.

When the sugar has dissolved, add the chopped apples and onions. Then add the mint leaves, bring to the boil and simmer for about 10 minutes. Throw in the raisins, add salt and pepper and simmer for another 5 minutes.

Spoon into warmed, sterilised jars, cover, and seal. This is great with Indian dishes.

Water Mint Ice Cream

SERVES 4

20 sprigs water mint, plus extra
 for serving
5 tbsp caster sugar
375g crème fraîche
3 egg whites

Wash the mint leaves thoroughly.
Put into a food processor or blender,
add the sugar and blend together.
Transfer to a bowl and stir in the
crème fraîche. Whisk the egg whites
to soft peaks, and fold carefully into
the mint mixture. Then, either put
into an ice-cream maker and follow
the manufacturer's instructions, or
put into a plastic container in the
freezer, stirring with a fork every half
hour or so to ensure that the mixture
doesn't crystallise. Serve with some
mint on the side.

Mint Water

I've included this simple recipe because mint-flavoured Turkish delight
is delicious, and you can substitute
mint water for rose water in the
recipe on page 166. Why not make
both kinds?

Pick as many mint leaves and stems
as you like, making sure that you cut
the stems rather than ripping up the
roots. Wash them thoroughly and
place in a large food-grade container
or crock. Pour boiling water over the
leaves, enough to cover them plus an
extra 2cm or so depth. Leave until
the water is cold, refrigerate overnight, then strain the water from the
mint. Store by freezing in bags or
clean plastic bottles (remember to
leave a space for the ice to expand).

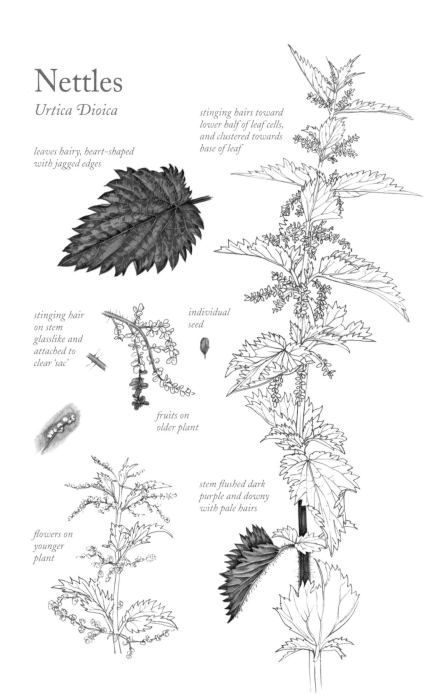

Nettles

Urtica Dioica

stinging hairs toward
lower half of leaf cells,
and clustered towards
base of leaf

leaves hairy, heart-shaped
with jagged edges

stinging hair
on stem
glasslike and
attached to
clear 'sac'

individual
seed

fruits on
older plant

stem flushed dark
purple and downy
with pale hairs

flowers on
younger
plant

Nettles grow all over the UK, quickly taking up residence in wasteland just about anywhere, but preferring damper soil. Nettles are happy to grow in full sun or in shade. They can reach nearly 2 metres in height, as anyone who has taken over a neglected garden will testify. The leaves are heart-shaped with jagged edges, hairy, and – this is the best-known way to identify them – the stems and leaves of the nettle sting like heck.

Culinary uses

The nettle's sting is completely removed by cooking, but if you're feeling strong-willed you can even eat nettles raw so long as you pound them well in a mortar and pestle. Although the sting is meant to be effective in counteracting rheumatism and arthritis, it's best to wear a decent pair of rubber gloves when harvesting nettles. Go for the tender young leaves of the first young shoots, which you can use exactly as you would spinach – in soups, in Greek-style filo parcels with feta cheese, or as a soup with a splash of water mint raita swirled in if you feel like going wild. Why not try them chopped into an omelette with wild garlic leaves, too?

Always use the young tops, because as well as tasting better than the older leaves, after the nettles flower and start to go to seed, the leaves can sometimes irritate the urinary tract.

Medicinal uses

Nettles are so abundant and so rich in protein and in vitamin C, it's surprising that we don't use them more often. They also help to increase our red blood cell count, thereby improving circulation.

Did you know?

During the First World War, nettle fibres were used to make the uniforms of the German army when cotton was scarce. And there's a particular kind of nettle that grows in the Himalayas that's used to weave a beautiful, lacy, dull golden-coloured fabric called aloo, which softens as it's washed.

Nettle Mushroom Crumble

SERVES 4

900g nettle tops
Olive oil, for frying
225g mushrooms, chopped
300ml white sauce
Sprinkle of nutmeg
100g breadcrumbs
50g mixed chopped nuts
2 garlic cloves, crushed
50g cheddar cheese, grated
Good knob of butter
Salt and pepper, to taste

Preheat oven to 180°C/gas mark 4.

Cook the nettles as you would spinach, in a saucepan with just enough water to cover them, then drain.

Heat a little olive oil and butter in a frying pan and sauté the chopped mushrooms, then remove from the heat and mix in the nettles and white sauce and add a sprinkle of nutmeg. Season with salt and pepper, and spoon into an ovenproof dish.

In a bowl, mix the rest of the ingredients together except the butter and spoon over the nettle mix, covering the top. Season again, dabbing the top with the butter, and bake for about half an hour or until the top is golden and bubbling.

Irish Nettle Beer

MAKES APPROX. 10 LITRES

12 litres water
About 100 nettle stalks
Juice of 1 orange and 1 lemon
1.5kg white granulated sugar
55g cream of tartar
7g sachet yeast (commercial dried bread yeast is fine)

Boil the water in a large pan and add the nettles, giving them a good stir. Cook for 15 minutes.

Turn the heat off and leave the mixture to infuse for 1 hour or until cooled to blood temperature.

Strain the liquid into another pan and discard the nettle sludge. Add the orange and lemon juice, sugar and cream of tartar. Heat gently until the sugar is all dissolved.

Remove from heat and leave to cool lukewarm, then stir in the yeast (follow manufacturer's instructions), cover and leave for 2–3 days.

Remove any scum from the surface. Then you can leave the liquid to ferment, as you would wine, or siphon into sterilised bottles. Leave for a month or so. The beer will fizz as it starts to ferment in the bottles so if you're using corks, make sure they are well secured.

Drink chilled with a sprig of fresh mint.

Nettle Syrup

The concept of nettle syrup might sound bizarre, but please do make this. You'll be glad you did, as it tastes great and is a good all-round tonic. Alternatively you can dilute with sparkling water and add a wedge of lemon or lime for a sophisticated summer drink that will keep your guests guessing. This recipe comes from gardening guru Lynne Allbutt.

MAKES APPROX. 2 LITRES

1kg young nettle tops
2 litres water
80g white granulated sugar for every
100ml strained liquid

Put the nettle tops and water in a large pan and bring to the boil. Simmer for 1 hour and then strain the liquid into a measuring jug, removing the nettles. Return the liquid to the pan and add the appropriate amount of sugar (see above). Simmer for another 30 minutes or so, until the liquid thickens and turns syrupy. Leave to cool, then siphon into sterilised bottles. The large amount of sugar will give the syrup a long shelf life, but you can also store in the freezer, either in small freezer bags or in an ice cube tray.

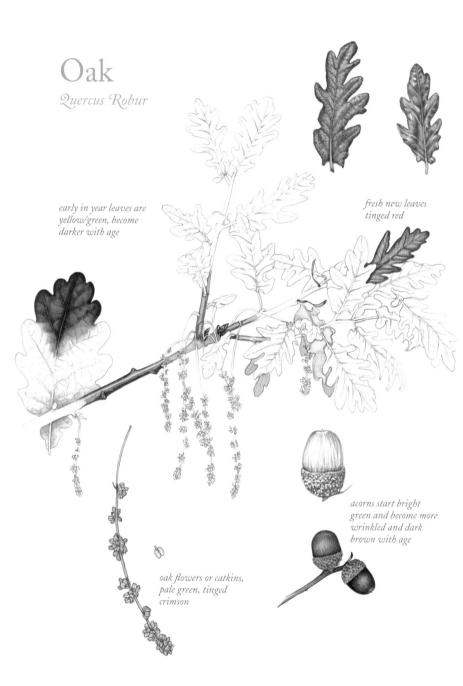

Oak
Quercus Robur

early in year leaves are
yellow/green, become
darker with age

fresh new leaves
tinged red

acorns start bright
green and become more
wrinkled and dark
brown with age

oak flowers or catkins,
pale green, tinged
crimson

There are several varieties of oak in Britain. The oak can be found in hedgerows or as a majestic stand-alone tree. The tree can grow to be very big (between 40 and 50 metres high) and very wide; the girth of the tree is a good indicator as to its age. The oak is deciduous, its leaves shaped in gently curving waves on either edge. The seed of the oak is, of course, the acorn, which is held in beautifully crafted little cups. The young, pale green-white acorn turns brown with age and ripeness. Sometimes you might see small balls stuck to the branches of the oak; these are popularly known as 'oak apples', made by the larvae of wasps.

The oak, because of its beauty, size, the great age it can achieve and the strength of its timber, is a symbol of strength and longevity, which is why it's one of the symbols of Great Britain.

Culinary uses
All parts of the oak can be used – leaves, bark, acorns and even the twigs, as you'll see from the following recipes.

Medicinal uses
A decoction of oak bark is good for throat infections and is also used to help break down kidney stones.

A poultice of cold oak leaves is good for burns and scorches.

In the Bach system of plant and flower remedies, oak is used for 'the plodder who keeps going beyond the point of exhaustion'.

Did you know?
The tannins contained inside oak mean that the wood is popularly used for making the barrels in which wines and whisky are matured.

Oak Leaf Liqueur

MAKES APPROX. I X 70CL BOTTLE

About 2 big handfuls fresh pale-green
  young oak leaves
70cl bottle white rum
225g granulated white sugar

Collect the oak leaves on a dry day,
strip from the twigs, wash and dry.
You'll need a sealable glass container
that's big enough to take all the
ingredients; simply pop all the ingredients into the container, seal, then
leave for a month before straining
the alcohol from the leaves into a
clean, sterilised bottle. Serve poured
over ice, or with soda water.

Acorn 'Coffee'

If you've access to a good quantity of
acorns, it's worth trying acorn 'coffee'
(although I hesitate to use the word
because roasted and ground acorns
really don't taste anything like coffee). Simply roast the acorns in their
shells (this makes the shells come
away more easily) in a slow oven for
about an hour, until they turn dark
brown, shaking them from time to
time to make sure the acorns don't
burn. Leave to cool before shelling
them and then grinding in a coffee
grinder.

   Alternatively, you can simply remove the shells and eat the roasted
acorns as they are, or, if you decide to
roast and grind them, they can be
used as an unusual flavouring in
sweet or savoury cakes and biscuits.

Oak Schnapps

MAKES I X 70CL BOTTLE

This drink has a fascinating taste that is hard to define until you're told what it is.

All you need to make it is a bottle of decent vodka. Oh, and an oak tree.

Use fresh, small oak branches and cut them with a small saw to about 2–3cm in diameter. Put about 20 of these, with the vodka, into a sealable glass container (or you can use the vodka bottle – that way there will be excess vodka to be taken care of!).

Then, all you need to do is put the container or bottle in a warm place for about a year, or until the liquid has turned a light browny-red colour. Fish out the wood, top up with more vodka if the taste is too strong, then leave for another 6 months or so, or until you can bear to wait no longer. This tastes great, served either at room temperature or chilled.

Ox-Eye Daisy
Leucanthemum Vulgare

*big flowers, soft white petals,
central florettes dark yellow*

*leaves smooth above, not clearly
veined, darker toothed edges,
paler below*

The dog daisy, moon daisy or ox-eye daisy will grow pretty much anywhere, although, like many wild flowers, it prefers poor, scrubby soils. They can grow quite tall – 1 metre or more – and the yellow-centred flower, with its halo of soft white petals, bobs on a thin wiry stem. The leaves are long, too, with several small coarse 'teeth' along either edge. The leaves at the base of the plant are more rounded.

Culinary uses

Although this section is about the ox-eye daisy – the large, beautiful flower that looks like a gigantic daisy and which frequently cheers the verges and hard shoulders of roads and motorways – the fresh, young leaves of the smaller daisies that grow close to the ground make a good addition to salads. You can also steam them or sauté them. They are worth a try because *everyone* knows what daisies look like, and you can find them just about anywhere in the UK.

Medicinal uses

Culpeper writes that the ox-eye daisy is 'a wound herb of good respect, often used in those drinks and salves that are for wounds, either inward or outward…' and that it is '…very fitting to be kept both in oils, ointments, plasters and syrups.'

Ox-eye daisy has similar qualities to chamomile and is used as an infusion to bring down fevers and night sweats. A tea of the leaves is effective against asthma. The flowers, steeped in boiling water and with honey added, soothe coughs and bronchial complaints. And a poultice of the pounded plant helps to heal external wounds, bruising and cuts.

Did you know?

Sometimes this daisy is called the maudlin daisy, or maudlinwort, not because it makes you feel miserable, but it is an abbreviation of St Mary Magdalene. Because the flower is useful in soothing women's complaints, it was dedicated to the goddess Artemis. This pre-Christian deity was replaced with a Christian saint, and the flower was renamed accordingly.

Dog Daisy Spread

MAKES APPROX. 250G

1 large carrot
1 small beetroot
1 medium onion
50g butter (or a good knob)
A good handful of fresh young daisy
  leaves, washed and allowed to dry
1 tsp powdered ginger
2 garlic cloves, crushed
1 tbsp plain flour
Juice of 1 lemon
Salt and pepper, to taste

Grate the carrot and beetroot and slice the onion into rings. In a pan fry the onion in the butter until it's soft. Add the grated carrot and beetroot and sauté for about 8 minutes. Add the daisy leaves and cook for another 5 minutes or so. Make sure nothing burns – you might need to add a little water.

Add the ginger and garlic and season with salt and pepper. Cook for another few minutes, then sprinkle the flour into the mixture, stirring all the while. Simmer for 5 minutes, then add the lemon juice. Turn off the heat and then, when the mixture has cooled slightly, blend with a hand blender.

This spread is lovely on toast or on jacket potatoes.

Pickled Dog Daisy Buds

MAKES 2 X 454G JARS

Approx. 1.5 litres (in volume) unopened
  ox-eye daisy buds
12 black peppercorns
8 allspice berries
2 pinches sea salt
12 mustard grains
1 garlic clove, finely sliced
800ml white wine vinegar

Wash the daisy buds and trim the stems. Leave aside to dry thoroughly. Take 2 x 454g jars and divide the black peppercorns, allspice berries, salt, mustard grains and garlic between them and then pack in the oxeye daisy flower buds.

Pour the white wine vinegar into a pan and bring to the boil. Take off the heat and allow to cool slightly, then pour the liquid over the contents of the jars, filling them to the brim. Secure with vinegar-proof lids, label, and store in a cool, dark, cupboard for at least 2 months to mature.

Chop for use in salads or pasta, or use as a replacement for capers.

Dog Daisy Raita

MAKES APPROX. 300ML

20g daisy leaves
150ml coconut milk
150g natural yoghurt
Juice of ½ lime

Wash the daisy leaves, then simply chop them and put into a bowl. Add the remaining ingredients and stir together. Serve at room temperature. Use the raita as you would 'normal' cucumber raita, as a cooling accompaniment to hot spicy dishes.

Dog Daisy Fritters

Cook these delicious tempura-style snacks in a deep-fat fryer, or in a wok, too, as long as you're careful. The oil needs to be very hot.

Use chilled, sparkling water to make your batter and don't worry if it's lumpy. Make the batter just before you use it – don't let it stand.

10 young daisy flowers, with stems
Oil, for frying

BATTER
85g plain flour
1 tbsp cornflour
½ tsp finely ground salt
200ml ice-cold sparkling water
A few chunks of ice cubes, crushed
Icing sugar, for dusting

Heat the oil. Sieve the flours and salt into a bowl. Whisk in the water and ice. One by one, hold the daisies by their stems and dip the flowers in the batter, swishing them about to coat thoroughly (don't worry about the petals sticking together in an unsightly lump), then drop them into the hot oil. The flowers will open up into a dramatic star shape, so allow enough room for them. When golden brown, lift out with a slotted spoon and dust with icing sugar.

Serve with cream, ice cream or panna cotta.

Pineapple Weed
Matricaria Discoidea

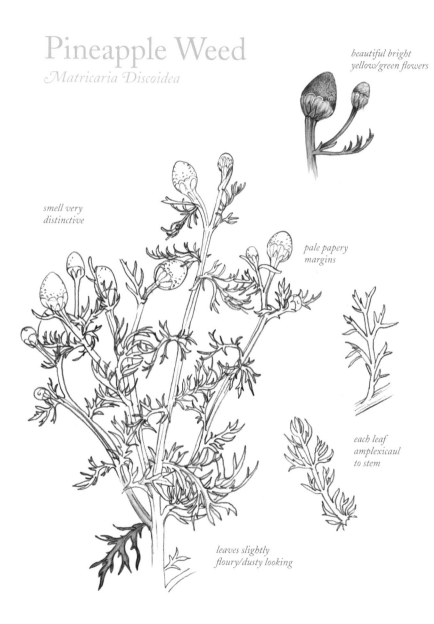

*beautiful bright
yellow/green flowers*

*smell very
distinctive*

*pale papery
margins*

*each leaf
amplexicaul
to stem*

*leaves slightly
floury/dusty looking*

This is one of those plants that many people remember from their childhood, perhaps from a time when we were literally closer to the ground and more liable to notice unusual plants. The pineapple weed is low-growing, often spreading along the ground, but can stretch to a height of 20cm. The leaves are feathery and the flowers seem to be all 'middle' and not much in the way of petals. These flowers, when crushed, have the most delicious scent that smells like a cross between chamomile and pineapple, hence the name.

Pineapple weed likes to grow in scrubby, marginal soil and, like dandelions, it thrives where it's walked upon, so you might see it along dirt tracks and the like.

Culinary uses
The flowers of pineapple weed make a refreshing snack, perhaps nibbled as you're taking a country stroll. You can use both the flower heads and the leaves in their raw state, thrown into salads, or added to sandwiches, for example. Or you can use to infuse a syrup, to make a pineapple weed cordial. A tea of pineappleweed is simple to make, too; simply take 3 or 4 flowers per cup, add boiling water, sweeten to taste and infuse for 4 minutes.

Medicinal uses
The infusion just mentioned can help settle an upset tummy and also has the effect of helping lower the temperature. Also, try putting a few fresh flowerheads in a bottle of water the next time you're out walking on a hot day. A poultice of the crushed plants is good for relieving skin complaints such as sores, ulcerations and blisters.

Did you know?
Pineapple weed started out on British shores as a garden flower. All the pineapple weed you see these days originally escaped from Kew Gardens in the 19th century. It's an effective insect repellent – if you're being attacked by insects such as midges, grab some pineapple weed and rub it into your skin, especially on your temples. You can also thread the flowers onto cotton and use to keep insects away from your clothes. Native Americans used to line their babies' beds with pineapple weed for the same reason.

MAKES APPROX. 1.5 LITRES

1.8 litres water
1.5kg granulated white sugar
Zest of 1 lemon
20 pineapple weed flowers

Boil together the water, sugar and lemon zest in a heavy-bottomed pan until the liquid has reduced by about a half. Take off the heat and add the pineapple weed flowers. Leave overnight and then strain the liquid from the solids.

Use as you would any cordial – add to sparkling water and serve over ice, or pour over yoghurt or ice cream.

(Ribwort) Plantain
Plantago Lanceolata

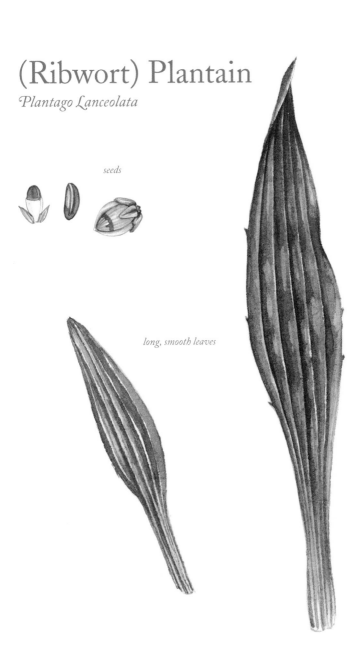

seeds

long, smooth leaves

There are quite a few different members of the plantain family (including sea plantain, buck's horn plantain, hoary plantain and greater plantain), but the one illustrated here is the ribwort plantain, which is readily available all over the UK. Though they share the same name, this leafy wild plantain of the UK is not to be confused with the banana-like vegetable that's used in many African dishes.

The plantain is a perennial plant, with long, smooth leaves and a short spike of flowers that appear in the summer on a long, tough stem. Plantains often take up residence in lawns, where irate garden purists use all sorts of hideous poisons to try to eradicate them. Best to live and let live, particularly since you can eat the plantain!

Do you remember as a child playing the game of 'guns'? It's probably the plantain flower that you used. You wind the stem around the flower head and pull it tight while tugging the knot forward, so that the flower head shoots off like a bullet.

Culinary uses

Plantain flowers have abundant seeds, which can be harvested and used in various dishes, both sweet and savoury. The seeds have an interesting nutty taste, which makes the effort of gathering them more than worthwhile. The young leaves of the plantain can be used as any green leaf; sautéed, steamed, even lightly roasted. The older leaves, though, are not worth bothering with because not only do they taste bitter, the long ribs have to be removed as they are unpleasantly tough.

Medicinal uses

If you have ever suffered from indigestion or irritable bowel syndrome you have probably bought expensive tubs of psyllium husks from a health food shop. These husks come from a member of the plantain family, *Plantago ovata*, which grows in India. The seeds of the British ribwort plantain, too, can be used to alleviate constipation and – conversely – the leaves are used to treat diarrhoea. If you want to try either of these remedies, you should contact a qualified medical herbalist.

Plantain leaves have been used for centuries in staunching bleeding and therefore helping wounds to heal. If you cut yourself and there are plantains close by, just put the leaf over the wound and hold firmly with the other hand.

Did you know?

Plantains are often spread via the soles of shoes and the tyres of cars; both can pick up the tiny seeds and take them on a journey elsewhere.

Ribwort Plantain Seed Pudding

An interesting and unusual recipe for plantain seeds.

SERVES 4–6

1 litre milk
4 thinly pared strips lemon zest
80g green ribwort plantain seeds
2 tbsp honey
Freshly grated zest of 1 lemon
2 tbsp lemon juice
80g caster sugar
3 egg yolks
Dollop of your favourite jam

Preheat oven to 180°C/gas mark 4.

Put the milk and the 4 strips of lemon zest in a pan and cook at a simmer for 2 minutes. Scoop out the peel, then pour in the plaintain seeds. Continue to cook, stirring, for half an hour, by which time the seeds will have started to thicken the milk. Take the pan off the heat and stir in the honey, grated lemon zest, lemon juice and sugar. Beat the egg yolks, combine with a little of the mixture from the pan, then add to the pan and stir vigorously for a couple of minutes.

Pour into an ovenproof dish and bake for 40 minutes. Remove from the oven and dot the jam over the top.

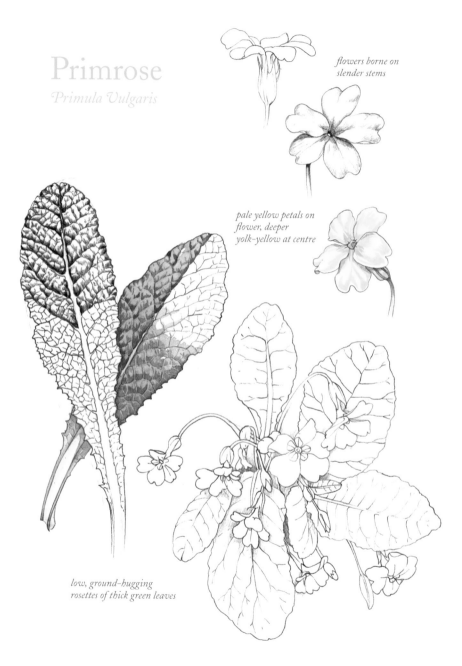

Primrose
Primula Vulgaris

flowers borne on
slender stems

pale yellow petals on
flower, deeper
yolk-yellow at centre

low, ground-hugging
rosettes of thick green leaves

The epitome of spring, the first little starry clumps of yellow primroses are a very welcome sight when they start to appear in April (19 April is officially 'Primrose Day', when a bunch of them is placed on Disraeli's grave in Buckinghamshire and also by his statue in Westminster Abbey – these were his favourite flowers).

The leaves form a rosette around the base of the flower. The flowers, which grow to about 10cm high, are pale yellow, each with 5 petals – although hybridised versions in all sorts of different colours are grown in gardens. They're often found at the edge of woods or in the grassy banks of hedgerows

Cowslips (*Primula veris*) are also known as primroses and, indeed, they do look similar. Cowslips are carried in a drooping bundle of flowers at the top of a stem, which can be quite tall. Sadly, neither cowslips nor primroses are seen quite as often as they used to be, so please be sparing when collecting them. Make sure you're picking them from a place where there are plenty of flowers, and don't, whatever you do, pull up the roots.

Culinary uses
Both kinds of primrose mentioned here are edible and look great in spring salads. The leaves, too, can be eaten and have quite a pleasant, sweet taste. Both the flowers and the leaves can be made into an infusion, or tea, which is said to calm the nerves. Also, the leaves can be cooked and used in the same way as any other green vegetable.

Medicinal uses
In terms of medicinal properties, the primrose is an incredibly useful plant. Primrose tea helps alleviate stress and anxiety, and both the root and the plant are used as an expectorant, helping to drive away nasty coughs and catarrh and easing conditions such as bronchitis. The plant has also been used as a remedy for rheumatism.

Did you know?
Not so long ago, primroses were so abundant in the countryside that they were picked, tied into posies, wrapped in tissue paper and packed carefully in boxes by rural families. The flowers were then sent by train to the big cities, to be sold at a premium.

One of the nicest ways of serving primroses is to crystallise the flowers. If you've never tried this, you really should. It's very easy and very effective. Do be sparing, though, since primroses look so pretty left growing in the wild!

1 egg white
A little orange blossom water or
  rose water
Caster sugar, for dipping
Primroses (flowers and leaves)

Beat the egg white with a little orange blossom water or rose water and paint it onto the primrose flowers and leaves with a paintbrush. Then dip carefully into the sugar, covering them well. Leave in a dry place overnight. Use the flowers to decorate cakes, cupcakes and chocolates. The measurements above should cover 3–4 average-sized flowers.

The sweet flavour of primrose leaves (not the flowers) combines nicely in this recipe, which dates from 1825. It first appeared in *The Mysteries of Trade or The Source of Great Wealth*, by David Beman. This is a treatise on chemistry and manufacturing – also containing 'receipts and patents'! This vinegar makes an intriguing addition to salad dressings. To make it, you'll need a food-grade plastic bucket with a sealable lid.

MAKES APPROX. 1.5 LITRES

1.7 litres water
300g brown sugar
500g tightly packed primrose leaves,
  washed
10g active yeast

Put the water and sugar into a heavy-bottomed pan and bring to the boil. Cook for 5 minutes, then add the primrose leaves. Take off the heat and, when the liquid reaches blood temperature, add the yeast. Cover with a cloth and leave in a warm spot overnight to allow the yeast to activate.

Pour into a food-grade bucket, cover loosely and allow to ferment. As soon as the fermentation stops (after about 10 days or so) put the lid on the bucket and set aside in a warm place. After a further 2 weeks, strain the vinegar into a sterilised jar and leave in a warm place to sour. After about a month your primrose vinegar will be ready to use.

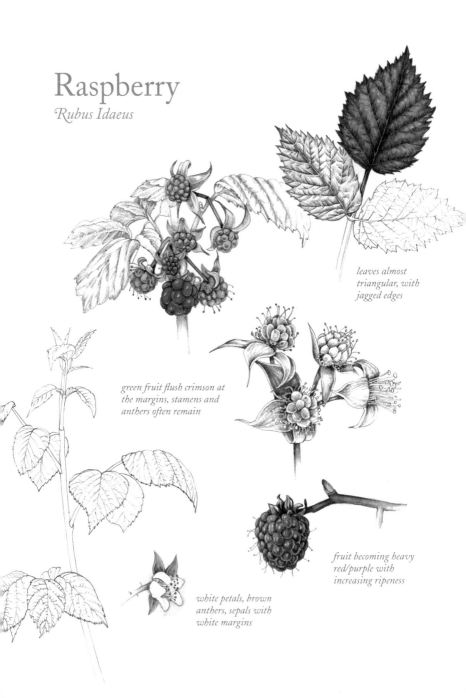

Raspberry
Rubus Idaeus

leaves almost
triangular, with
jagged edges

green fruit flush crimson at
the margins, stamens and
anthers often remain

fruit becoming heavy
red/purple with
increasing ripeness

white petals, brown
anthers, sepals with
white margins

The wild raspberry is possibly my favourite ingredient in this entire book. Wild raspberries are small, but delicious, and it's a real treat when their sparkling jewel-like colours start to gleam out from the hedgerows. And raspberries are incredibly versatile; there are so many things you can do with them. You can make jams and jellies, chutneys, pies, ice cream, syrups and sorbets, all from this one delicious fruit.

A perennial plant, raspberry has whippy, woody stems with small thorns – although they are not really big enough to be called thorns. The leaves are almost triangular in shape, with jagged edges. The white flowers appear from May onwards, followed by those lovely red berries from June to October.

Culinary uses

There's nothing better than eating wild raspberries just as they are, or stirred into yoghurt. But if you do have an abundance of them, try adding them to a cupcake mix or make them into jam. One of the loveliest home-made wines is made from raspberries, and the recipe is included here.

Medicinal uses

Raspberry leaves were used for centuries as an infusion to help women in childbirth during contractions – in fact, the remedy is so old that its use in this way is recorded by the ancient Greeks. Then, a couple of millennia later, commercially available drugs took over. During the Second World War, however, a shortage of these manufactured drugs meant that midwives reverted back to the old remedy and it was discovered that the use of raspberry leaf tea to ease childbirth is no old wives' tale, but really does work. The leaves contain a chemical called fragarine that can both ease and stimulate the uterine muscles. Raspberry leaves can also alleviate menstrual pain. The fruit is mildly laxative.

Did you know?

Raspberry varieties also come in black, purple and even gold and yellow.

Raspberry Leaf Tea

Making raspberry leaf tea is as simple as putting the leaves in a pot and adding hot water and honey or sugar, and leaving to steep for about 5 minutes.

Raspberry Wine

To make this wine, you will need a large food-grade bucket with a lid, a demijohn with airlock, and some sterilised bottles with corks.

MAKES 3 X 70CL BOTTLES

750g wild raspberries, washed
3 litres boiling water, plus 1.5 litres
Juice of 2 oranges
1kg granulated white sugar
1 x 7g sachet wine yeast, made up
  according to manufacturer's
  instructions

Put the fruit into a large food-grade bucket with the 3 litres boiling water. Smash the fruit with a plastic spoon, cover, and leave for 2 days.

Strain the juice from the solids through muslin and add the orange juice.

In a heavy-bottomed pan, boil the sugar with the rest of the water and add to the raspberry juice. Leave to cool to blood temperature and then add the yeast. Cover again and leave to ferment for 3 days.

Siphon into a demijohn and seal with an airlock. Leave again to ferment for about 3 months and then siphon into sterilised bottles, being careful not to disturb the sediment (or 'lees', as it is called) at the bottom of the demijohn. Cork, and store in a cool dark place. Tie the corks on with string, since the wine will continue to ferment in the bottles. Leave for 1 year; if you're lucky you will find that the wine may be sparkling when you serve it.

Raspberry Vodka

This is delicious and so simple to make. You can use any soft fruits to make different-flavoured vodkas. In addition to your full bottle of vodka, you will need a clean, empty bottle.

Enough raspberries to fill 1 x 70cl bottle
70cl bottle good-quality vodka
Sugar (optional)

Wash the raspberries and dry them. Pour half of the full bottle of vodka into the empty bottle. Top both bottles up with the raspberries. Keep for 3 months in a cool dark place, remembering to shake the bottles occasionally.

After 3 months, pour the vodka away from the fruit. You can add sugar, if you like, although frankly I don't think the recipe needs it. And don't throw the alcohol-infused raspberries away – eat them with ice cream.

Raspberry Vacherin

The combination of fresh wild raspberries, sweet sticky meringue and the sharp tang of crème fraîche makes this a sensational dessert.

SERVES 6

MERINGUE
2 egg whites
100g caster sugar
2 drops vanilla essence
A few drops lemon juice

Preheat oven to 130°C/gas mark ½. Line a baking tray with parchment paper.

Whisk the egg whites in a squeaky-clean bowl, until they form stiff peaks. Beat in the sugar, vanilla and lemon juice.

Spoon 12 blobs of the mixture on the baking tray and bake in the cool oven for 1 hour. Turn the oven off, but leave the meringues in the oven overnight.

FILLING
225g wild raspberries, plus extra
  to garnish
300ml crème fraîche
2 tbsp caster sugar
1 egg white
A few mint leaves, to garnish

Mash half the raspberries, leaving the other half whole, and stir all the fruit into the crème fraîche. Fold in the sugar. Whisk the egg white to soft peaks and fold into the crème fraîche mix.

Make 'sandwiches' of the meringues and the raspberry mix. Serve with a few raspberries on the side, plus some mint leaves for garnish.

Wild Raspberry and Meadowsweet Jam

This recipe is for the best jam I've ever tasted. Also called Angus Hedgerow Jam, it's from Fi Bird of Stirrin' Stuff, an organisation she started with her husband to stimulate childrens' creativity as well as educating them about maths, chemistry and biology and using cooking as a fun common denominator.

MAKES 6 X 454G JAM JARS

1.3kg wild raspberries
285ml water
1.3kg golden granulated sugar
4 sprigs (flowers) meadowsweet
Knob of butter

Wash the raspberries and put them in a heavy-bottomed pan with the water and sugar. Wrap and tie the meadowsweet in muslin and put it with the fruit. Dissolve the raspberries and sugar over a low heat and then increase to a rolling boil for 15 minutes, adding the knob of butter to dissolve any scum that might form on the top.

Test to make sure the jam will set – put a blob on a chilled saucer and, if it's 'done', the surface of the blob of jam will wrinkle when you push it with your finger.

Remove the muslin bag of meadowsweet and allow the fruit to settle before pouring into warmed, sterilised jars to set. Cover with wax discs and seal the jars. Store in a cool, dark place.

Red Clover

Trifolium Pratense

fragrant deep magenta pink to red flowers, round to oval shaped heads

older flowers tinged reddish-brown

trifoliate leaves, mid-green with paler green or white v-shaped marking

A low-growing perennial herb occupying grassy meadows and verges, the clover flower is a deep pink/purple/red colour, like a pompom of many tiny flowers. The distinctive trifoliate leaves are borne at the end of a stem.

Clover is one of those plants, like many others in this book, that are easy to take for granted. But look closely at those pinkish-red flowers; they're absolutely beautiful. I am convinced that if they were difficult-to-grow hothouse flowers, red clover blossoms would be worth a fortune. Children in the know still suck the flowers, their sweetness released by the heat of the sun, to taste that scented sticky juiciness.

As a crop, clover is incredibly useful in that it enriches the soil, making the grass more nutritious for the cattle that feed on it.

Culinary uses

Luckily, both flowers and leaves are edible for humans. Red clover flowers look and taste lovely in salads – be sure to shake out any tiny insects that might be hiding in amongst the many folds and petals. As well as adding the whole flowers, you might persuade people to eat the clover by pulling the individual flowers away from the flower heads, too.

Try adding individual clover flowers, with a little sugar and salt, to rice. Throw the flowers in after the rice is cooked but still hot.

Clover leaves can be sautéed with garlic and onion and used as any other green vegetable, eaten raw in salads, steamed or added to pasta.

Medicinal uses

Red clover leaves, like willow bark and meadowsweet, contain a small amount of salicylic acid, the main constituent of aspirin. Therefore, an infusion of them can alleviate the symptoms of headaches, painful menstruation, and other aches and pains. The leaves also contain calcium and protein.

The flower heads can be applied externally for a number of skin complaints, including burns and eczema. They are also sometimes used as an ingredient in throat medicines since they have a soothing effect.

Did you know?

A four-leafed clover is, of course, universally acknowledged as a lucky charm, but a two-leafed clover is connected with marriage. According to an old tradition, should an unmarried girl find a two-leafed clover, she should put it in her shoe. She will then marry the first young man that she meets – or, failing that, another young man with the same name!

Red Clover Lemonade

This not only tastes good, but *is* so good for you.

MAKES APPROX. I LITRE

750g fresh red clover blossoms, plus extra to garnish
500ml water
500g honey (or sugar if you prefer)
Juice of 2 lemons

Simmer the clover blossoms in the water, in a covered pan, for 10 minutes. No need to boil. Then add the honey or sugar and stir until it's dissolved.

Cover and let the mixture steep and cool for several hours or overnight. Steeping makes the infusion strong, increasing the potency of the calcium and other nutrients in the clover.

Finally, add the lemon juice and chill in the fridge.

Strain, then serve poured into tall glasses over ice, garnished with a chunk of lemon and a couple of clover flowers.

Red Clover Almond Biscuits

These make a great accompaniment to the red clover lemonade.

MAKES APPROX. 25 BISCUITS

380g wholemeal flour
3 tsp baking powder
100g almonds
100g butter
2 eggs
120ml buttermilk
¼ tsp almond extract
287g red clover flowers, plucked out of the flower head

Preheat oven to 230°C/gas mark 8.

Whizz the flour, baking powder and almonds in a food processor until everything is finely chopped.

Add the butter and pulse until you have a crumbly texture. You may need to add some extra butter.

Add the eggs, buttermilk, almond extract and the red clover flowers. Whizz once more until you have a lump of dough.

Roll out the dough (not too thin) on a lightly floured work surface. Then cut into 5cm squares. Transfer to an ungreased baking tray and bake for 10–15 minutes, or until the biscuits are golden brown.

You can serve the biscuits on their own or with a little butter.

(Dog) Rose

Rosa Canina

long, fine, thorned
branches

cross section

colour of rosehips
from orange to
crimson

rosehips withering

individual seed,
hairy

In its wild form, the most common rose in Britain is the lovely dog rose, which usually has pale pink flowers, followed by shiny red hips in autumn. The deciduous dog rose is often seen in hedgerows, its long, fine, thorned branches arcing through other plants. It flowers throughout June and July. The leaves have two to three pairs of symmetrically-spaced leaflets borne along a stem, and the thorny arching branches are a hazard.

Culinary uses
There's so much that you can do with the rose that I've divided the recipes into two sections: rose petals and rose hips.

Medicinal uses
Rose hips are incredibly rich in vitamin C – containing 20 times as much as oranges. The seeds in those hips are a diuretic and the leaves can have a laxative effect.

Did you know?
It's likely that the dog rose was originally named the 'dag' rose, after its dagger-like thorns.

The hairs inside the rose hips can prove irritating to the throat, so they're usually removed before cooking. However, children have been known to use these miniscule hairs as itching powder!

ROSE PETALS

The flavour of rose petals is *so* evocative of Eastern glamour and so delicious, I'm surprised we don't use them more as a flavouring in the UK, particularly as we're so fond of roses and they're so prolific.

Rose Petal Yoghurt

This is outrageously delicious, the sort of dish that you would imagine being brought to an Arabian princess on a golden platter. And yet three of the ingredients are common British wild plants.

The exotic flavour of this dish belies the simplicity of making it. Choose a highly scented variety of rose. Wild dog roses are perfect, as are *Rosa rugosa*, the ubiquitous shrub used for hedging, which generally has bright pink or white flowers. Remember to pick the rose petals when the full sun is on them.

SERVES 2–3

*Good handful of dried rose petals
  (see below)*
Honey, to taste (optional)
1 large tub thick Greek yoghurt
Small handful of wild strawberries

TO SERVE
*Couple of generous sprigs mint (water
  mint, garden mint or calamint are all
  fine), chopped*
Pistachio nuts, chopped

To dry the rose petals, simply spread them out on sheets of absorbent paper (but not newspaper) in a warm, dry place. This should take a couple of days.

If you're using honey, stir it into the yoghurt thoroughly. Then add the crumbled dried rose petals and the wild strawberries.

Leave in a cool place for 2–3 hours to allow the rose flavour to infuse, and serve topped with the chopped mint and pistachio nuts.

Rose Petal Syrup

Another good staple way to use rose petals, this syrup is a handy ingredient for other rose recipes. I have used American 'cups' to measure the ingredients since it's difficult to weigh fresh rose petals on scales. You can make as much or as little as you like, keeping to the proportions used below, but it has such a lovely flavour that I recommend you make a large quantity. Its high sugar content means it will keep well in sealed jars, like jam.

1 cup rose petals
1 cup water
1½ cups granulated white sugar
1 star anise (or a couple of shards
  cinnamon stick or 2 cloves)

Put the rose petals into a pan with the water and bring to the boil. Simmer for 4–5 minutes. Add the sugar and spice and simmer again, slowly, to dissolve the sugar. Let the syrup cool a little before straining it. Store in a sealed jar if not using straight away.

Rose Water

It's surprising that rose water isn't found more often in peoples' larders or store cupboards. It imparts an exotic touch to cakes, pastries, and even glazes. And if you *really* want to impress people and have a glorious afternoon to boot, you can use it to make your own Turkish delight (see page 168).

Collect as many scented dog rose petals as you can find. Remember to pick them when they are warm from the sun, and dry. Take a large pot or food-grade container (anything that's large enough and that won't contaminate the delicate flavour of the rose petals) and put the petals in it. Pour boiling water over them, making sure that they are covered, and then add another 2cm or so.

Let the mixture stand until the water is cool, then put it in the fridge overnight. If the container you have used is too large for the fridge, leave in the coldest place possible.

The next day, strain the liquid from the petals. That's it – your own rose water.

The best way to store your rose water is to freeze it. Use plastic water bottles, freezer bags, or freeze in ice cube trays.

Rose Petal Sorbet

SERVES 3–4

230g granulated white sugar
570ml water
*4 large handfuls fresh rose petals, picked
  in full sun*
Juice and peel of 1 lemon
2 tsp glycerine

Boil the sugar and the water in a
heavy-bottomed pan for 5 minutes.
Take the pan off the heat and add
the rose petals and then lemon peel
and juice. Mash and leave the mixture to stand overnight, covered with
a clean muslin cloth or tea towel so
that the steam doesn't condense on
the inside of the pan.

The next day, strain the liquid from
the rose petals. Add the glycerine,
then either put into an ice-cream
maker and follow the manufacturer's
instructions, or simply put into a tub
and freeze for 2–3 hours, stirring
occasionally so that the mixture
doesn't crystallise.

Rose Petal Jam

MAKES 2 X 454G JARS

450g rose petals, picked in full sun
900g granulated white sugar
2.2 litres water
Juice of 4 lemons

Put the clean, dry petals in a large
bowl with half the sugar. Stir to mix,
then cover and leave in a warm place
overnight.

Take a large, heavy-bottomed pan
and add the rest of the sugar, the
water and the lemon juice. On a
low heat, stir until the sugar has
dissolved, then stir in the rose petals
and sugar mix. Simmer for about 20
minutes, then turn up the heat and
boil for another 5 minutes, or until
it reaches setting point. Test by
putting a blob on a chilled saucer;
if it wrinkles when you push it with
your finger, it's ready. If it's not ready,
boil again until the setting point
is reached.

Transfer to warmed, sterilised jars,
cover with wax discs, and seal. Store
in a cool place.

Rose Turkish Delight

A perfect way to use your delicious home-made rose water. This recipe calls for scalding hot ingredients, which means it isn't safe for children to make unsupervised. You'll also need a sugar thermometer.

MAKES APPROX. I KG

760g granulated white sugar
2 tsp lemon juice
350ml water, plus 650ml
190g cornflour
1 tsp cream of tartar
2–3 drops red food colouring
1½ tbsp rose water
130g icing sugar

Line a 22cm square baking tray with a double layer of strong foil and spray the foil with non-stick cooking spray. Don't be tempted to use butter since this will contaminate the flavour. Make sure the foil comes up the sides of the tray since you will use the foil later as a handle.

Put the sugar, lemon juice and the 350ml water in a heavy-bottomed saucepan over a medium heat. Stir until the sugar dissolves, then bring the mixture to the boil. Brush down the sides of the pan with a wet pastry brush to prevent sugar crystals from forming, and insert a sugar thermometer.

Let the mixture boil, without stirring, until it reaches 120°C on the sugar thermometer. Keep brushing the sugar down the sides of the pan.

When the sugar syrup is around 110°C, you can begin to prepare the rest of the ingredients (keep the sugar syrup boiling, though). Place the 650ml water in another, slightly larger, saucepan. Add the cornflour and cream of tartar and whisk until the cornflour dissolves and there are no lumps. Place the saucepan over a medium heat and bring the mixture to the boil, stirring or whisking it constantly. The mixture will become thick and pasty.

Once the sugar syrup is at 120°C, remove it from the heat. Slowly pour it into the cornflour mixture, whisking until it is fully incorporated. Be careful, since the liquid will be very hot.

Reduce the heat to low and simmer, whisking it every 8–10 minutes, for about 1 hour, until the candy has turned a light golden-yellow colour and is very thick.

After 1 hour, remove from the heat and stir in the food colouring (not too much unless you want the Turkish delight to be blood red!) and the rose water. Pour the candy into the prepared baking tray and allow it to set at room temperature, uncovered so that it doesn't condense, overnight.

The next day, remove the Turkish delight from the pan, using the foil as handles. Dust a work surface with icing sugar, and flip the Turkish delight onto it. Remove the foil from the back and dust the top with more sugar. Use an oiled chef's knife to cut it into small squares. Dust each side of the squares with more icing sugar to prevent stickiness.

Turkish delight is best eaten fresh, soon after it's made. It will keep, if stored carefully. To do this, put it in a container that's been lined with waxed paper, and put more waxed paper between the layers of sweets. Revive before serving by dusting with more icing sugar.

Gulkand

This is an unusual Indian recipe but one that is very easy to make. It's used in Ayurvedic medicine to soothe and cool a 'hot' temperament. Medicinal use notwithstanding, it's delicious.

All you need to do is take equal quantities of rose petals and granulated white sugar, layer them in a jar, and seal. The Indian recipe then calls for the sealed jar to be 'cooked' for up to 6 hours a day in the full heat of the sunshine until the petals have turned brown and released their juice into the jam. Even in India this can take a few weeks! I have found, however, that British sunshine doesn't tend to stick around for long enough to do much consistent 'cooking', and so the mixture needs a little helping hand.

Put the sealed jar in a pan of warm water for a few hours every day, until the petals are brown and the sugar has absorbed the fragrance of the petals.

The texture of the Gulkand is rather like a grainy fudge. You can eat it on its own, or with ice cream, or added to sundaes and sorbets.

ROSE HIPS

What a versatile flower is the rose! Beautifully scented, prolific … and especially generous. Not only can we use the petals, but after the flowers come those fabulous hips, chock full of vitamin C and very plentiful. The hips appear just as the summer days are shortening to autumn, providing a joyful splash of flame colour in the hedgerows.

Rose Hip Syrup

There's a very good reason why old remedies are best; it's because they work. Rose hip syrup has been used as a cure for the common cold for generations and so it seemed a good idea to include it here. It's not only good for you, but it has such a gorgeous taste that no one would ever consider it a medicine – a good excuse to make plenty of it. Freeze it and let the kids enjoy it diluted in hot water or glooped over ice cream, or maybe even with a dollop of cream in their porridge.

You will need a large food-grade container, a jelly bag for straining the syrup and some sterilised bottles or freezer bags.

MAKES APPROX. 2.5 LITRES

2 litres water, plus 1 litre
1kg rose hips
450g granulated white sugar

Boil the 2 litres water in a large pan. While the water is heating, prepare the rose hips. After making sure they are of good quality, remove the stalks and mince the rose hips (do this as soon as you can after picking). Once the water is boiling, throw in the rose hips. Bring back to the boil, then take off the heat and leave for 15 minutes.

After 15 minutes, let the juice drip through a jelly bag into a large food-grade container. Don't throw away the pulp – put it back in the pan with the 1 litre (boiling) water, bring to the boil again, leave for 10 minutes, then strain once more.

Pour the strained juice into a clean pan and reduce until you have 1 litre left in the pan – gauge this by pouring 1 litre of cold water into the pan before you start the process and make a mental note of the water mark in the pan.

Once the juice has reduced, add the sugar and stir until it has dissolved. You can then funnel the syrup into warmed, sterilised bottles while the liquid is still hot, or if you decide to store it in plastic freezer bags, allow it to cool first.

Rose Hip Jus

MAKES APPROX. ½ LITRE

1kg rose hips
50g soft dark brown sugar
Water (see recipe for amount)

First, you need to split all the rose hips in half and scrape out the seeds – enlist some volunteers! Why remove the seeds? Well, the little hairs you find inside the pods are an irritation for most people.

Then all you do is put the prepared hips, the sugar, and the same volume of water in a large, heavy-bottomed pan, bring to the boil, then simmer until the hips are soft. The easiest way to determine the volume of water is to see how far up the pan the rose hips come, and add the same depth of water. Leave to cool, then pour into a blender (or use a hand blender) and blend until smooth.

This jus goes well with hot and cold savoury dishes, and cheese.

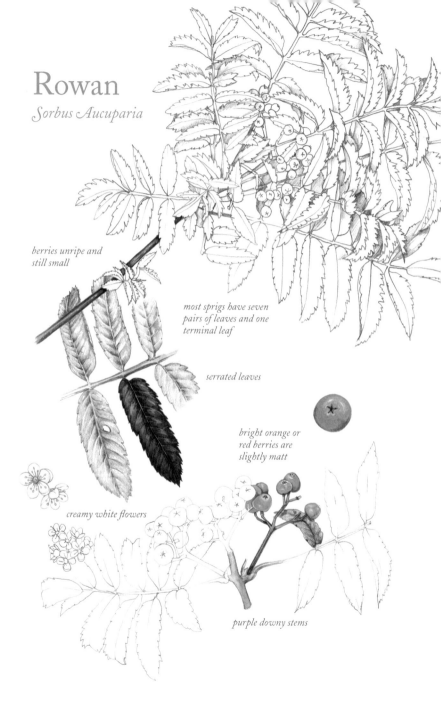

Rowan
Sorbus Aucuparia

berries unripe and
still small

most sprigs have seven
pairs of leaves and one
terminal leaf

serrated leaves

bright orange or
red berries are
slightly matt

creamy white flowers

purple downy stems

Although the rowan tree is also called the mountain ash, *Sorbus aucuparia* isn't actually a member of the ash family.

This is a small tree, no more than 20 metres tall, and not generally found in traditional hedgerows but more as a distinctive stand-alone tree, particularly in hilly areas. The leaves are a little like a larger version of rose leaves, and the tree is actually a member of the same family as the rose. The blossoms of the rowan are sometimes confused with those of the elderflower, but the rowan's panicles are tighter and more 'controlled' looking. The blossom appears from May onwards and the bright orange-red berries start in August.

In folklore, the rowan is one of many trees in Britain that is said to have connections with the magical world. A cross made of rowan wood, and tied with a red thread, was considered to protect against witches, but, conversely, the tree itself was said to grow where witches lived. Not surprising, really, since the tree often appears in isolated places. One of the folk names of the rowan is 'witchwood'.

Wine made from rowan berries is said to confer the gift of second sight to the drinker; there's only one way to find out if this is true – by making the rowan berry wine described here.

Culinary uses

Rowanberries are very versatile; you can transform them into wine, jellies, and syrups. They must always be cooked however, as the raw berries can cause an upset stomach and vomiting.

Medicinal uses

Rowan berries are a good source of vitamins C and A, and the fresh juice can be used as a gargle to relieve sore throats. Rowan berry jam can help ease diarrhoea, while an infusion of the berries benefits haemorrhoids.

Did you know?

As the rowan was once renowned as a tree of magical protection, even cream was stirred with rowan twigs to stop any lurking witches from curdling it!

Rowan Berry Wine

You will need a large food-grade container with lid, a couple of demijohns with airlocks and some sterilised bottles with corks.

MAKES 3 X 70CL BOTTLES

*1kg ripe rowan berries, washed, dried
  and with stalks removed
1.3kg granulated white sugar
2 litres boiling water
Juice of 2 oranges or lemons
7g sachet wine yeast, made up according
  to manufacturer's instructions*

Put the berries in a large food-grade tub, one that has a lid.

Boil the sugar and water in a large, heavy-bottomed pan and then pour over the berries. Mash thoroughly, add the orange and lemon juices and, when the mixture has cooled to blood temperature, add the yeast. Leave for 3 days, stirring occasionally.

After 3 days, the mixture will smell appalling; don't worry. Strain the juice into a demijohn, avoiding any sediment at the bottom of the tub. Top up with fresh cooled boiled water if necessary. Seal with an airlock and leave for 3 months, then strain into a second demijohn. Leave for another month and then siphon into sterilised bottles, remembering to tie the corks down since the fermentation process will continue in the bottle. Store in a cool, dark place for about 1 year before drinking.

Rowan Jelly

You will need a jelly bag for straining the juice, or use several layers of cheesecloth.

MAKES APPROX. 2KG

*900g apples, peeled, cored and sliced
(you can use crab apples if you like)
1 litre water
1.3kg rowan berries, washed, dried and
stalks removed
Soft brown sugar (see recipe for
quantity)*

Put the sliced apples and the water in a pan and bring to the boil. Boil until the apples soft.

Add the rowan berries and simmer until the whole thing is a pulpy mash. Strain the juice through a jelly bag, allowing it to drip for as long as necessary (maybe overnight).

Measure the juice: you will need to add 450g sugar for every 600ml juice. Warm the sugar first in an oven, then boil the juice in a large pan, add the warmed sugar and boil again for 10 minutes. You'll need to keep skimming the scum from the top of the pan.

Pour the mixture into warmed, sterilised jars, or alternatively pour into a shallow tray so you can cut into cubes and serve like Turkish delight.

Rowan Gin Or Vodka

This vodka takes a couple of months to mature, so make as soon as the berries are ripe if you want to give this delicious aperitif as a Christmas gift. Freeze the berries before using to simulate the action of frost – this helps the fruit to give up its juice and is easier and less messy than the alternative method of pricking the fruit all over with a pin.

*Enough rowan berries to fill 1 x 70cl
bottle
70cl bottle gin or vodka*

Pick over the berries and remove any stems. Wash thoroughly, and leave to drain before spreading on a tray and setting in the sun to dry. When completely dry, put into freezer bags and place in the freezer for at least a week.

After this time, place the frozen fruit in a large jar and cover with the gin or vodka. Seal tightly, then place in a dark place to infuse for at least 4 weeks. Each time you pass by, agitate the jar to help the infusion process.

Decant the liquid from the solids, and filter again into bottles. Seal tightly and place in a cool, dark cupboard to mature for at least 8 weeks before serving.

Samphire
Salicornia Europea

*fleshy stems, succulent,
bright green, like
asparagus in miniature*

*fleshy, red and dusty,
taste very salty*

This seashore plant has narrow segmented stalks, with little lobes between the main stem and the branch, sometimes with tiny yellow flowers at the junction of those lobes.

The elegantly named marsh samphire is also known as glasswort, picklewort or sea asparagus. In order to grow it needs not only water, but also large quantities of salt. Naturally then, samphire grows well near the sea, along the edges of beaches and in salt marshes. Rock samphire, or *Crithmum maritimum*, can be used in the same way, but tends to be harder to find and, even if you find it, more difficult to get at, since it favours taking root in hard-to-reach crevices that can be dangerous to access. In fact it's mentioned in King Lear:

'... *Stand still. How fearful*
And dizzy 'tis to cast one's eyes so low!
The crows and choughs that wing the
 midway air
Show scarce as gross as beetles. Halfway
 down
Hangs one that gathers samphire,
 dreadful trade!
Methinks he seems no bigger than his
 head ...'

Culinary uses
Samphire, which was previously only really used to decorate fishmongers' stalls and window displays and often simply thrown away at the end of the day, has become something of a foodie cult as we've discovered just how delicious, as well as nutritious, it is. The taste? Of the sea ...

The best time to gather samphire is when the shoots are young, round about May. Cut the stems, rather than tugging at them, so that you don't damage the roots.

Medicinal uses
Samphire isn't really used in any medical applications, although it does contain a large amount of mineral salts and has a mild diuretic effect. In times gone by, the plant was believed to be a cure for headaches. Like most seaweeds, both samphires are rich in minerals.

Did you know?
Samphire's old folk name of 'glasswort' sounds incongruous but there's a very good reason for it. It is a key source of sodium sulphate, which was used in the manufacture of glass until the 19th century.

Samphire with Garlic and Lemon Juice

SERVES 4–6 AS A SIDE-DISH

2 litres water
500g marsh samphire
4–6 garlic cloves, crushed
2 tbsp lemon juice
3 tbsp extra-virgin olive oil

Put the water into a saucepan and place on a high heat (there's no need to add salt). While the water is heating, cut off the roots of the plant, separate the individual strands and wash thoroughly under cold running water (samphire contains quite a bit of sand).

Once the water is boiling, reduce the temperature slightly, add the samphire and leave for a few minutes – long enough to blanch, but you don't want to overcook it. Drain off the water and let the samphire cool.

Mix the garlic with the lemon juice and oil. Once the samphire has cooled, brush the smaller side branches, which are tender, from the main stalks (which generally are a bit tough) by holding the main stalk near the root and rubbing your hand along it lightly. You can eat the main stalks, by the way, by dragging them through your teeth but it's easier to eat if you present them – main stalks and tender side shoots – separately.

Place the side branches in a salad bowl. If the main branches are not overly tough (try one), add them as well and, finally, drizzle the garlicky lemon dressing over them.

Samphire Pickle

Sharp and tangy, this is a really good way of using an abundance of samphire. It's gorgeous piled onto warm crusty bread along with some strong cheddar cheese or melting Brie.

MAKES APPROX. 450G

500g samphire, washed thoroughly
50g sea salt
600ml water
Pickling vinegar (see next recipe for a fast way to make this)

Soak the samphire sprigs in a bowl of sea salt and water for 24 hours. Rinse well and boil in pickling vinegar for 7–10 minutes.

Transfer the samphire and vinegar to warmed, sterilised jars and seal tightly after the liquid has cooled. Keep for 3 weeks before using.

Fast Spiced Pickling Vinegar

MAKES I LITRE

Approx. 7g each of the following whole spices (i.e. not ground): Cloves, cinnamon, white peppercorns, allspice, ginger
1 litre vinegar

Take a clean tea towel, put the spices on it and fold the towel over them (you can add more spice if you prefer a spicier pickle). Thump the spices with a rolling pin.

Add the spices to the vinegar in sterilised jars, and seal. Leave for a couple of months, shaking the jars daily. After 2 months, strain the vinegar from the spices.

If you can't wait 2 months for your vinegar to absorb the flavours of the spices, simply put all the ingredients into a bain-marie (with a lid). Making sure the vinegar is cold to start, bring to the boil slowly, then take the pan off the heat and leave for 2 hours (or until the liquid is cold) with the lid on. Then scoop the spices from the vinegar.

Sloe
Prunus Spinosa

*long leaves at branch tip,
further down the sprig rosettes
of shorter leaves with fruit
amongst them*

*orange anthers, pure white petals
with green centres, petals tinged
blueish purple*

*first blooms of the year
highly scented*

The bush that bears the sloe is called the 'blackthorn' and this is an apt name, because the shrubby tree (which can reach a height of 4 metres) has very spiny thorns and shiny black berries. The leaves are small and circular, but by the time the sloes appear they've usually fallen, making the sloes very easy to spot once you get the 'eye' for them. Tiny white blossoms appear on the tree from May onwards.

The little black sloes, which are about 5mm–1cm long, are the great-grandmothers of our cultivated garden plums. These fruits are preceded by little white flowers that appear from May onwards.

Culinary uses

The flowers are edible and taste a bit like almonds – use them as an unexpected addition to salads, frozen into ice cubes, or even as a cake decoration. You can also make a syrup of them.

Medicinal uses

Sloes contain vitamin C. A tea made of the flowers has both a tonic and laxative effect. A tea made of the leaves, with added sugar or honey, soothes throat ailments such as laryngitis and tonsillitis.

Did you know?

When gathering blackthorns, take a small sachet of antiseptic wipes with you. Wounds from those ultra-sharp thorns are painful and can become infected.

Blackthorn Blossom Syrup

MAKES APPROX. 5 LITRES

Simply heat 450g granulated white sugar in 4.5 litres water, adding a couple of big fistfuls of the flowers, simmer until the sugar has dissolved, and leave to cool. Pour into sterilised jars or bottles, and seal.

Use on desserts and ice cream, or add to sparkling white wine to fabulous effect.

Sloe Gin

Undoubtedly, the best-known way of using sloes is to make sloe gin. Most recipes ask you to prick the skins of the fruits with a needle, but this is purely to emulate the action brought about by frost. When the juice inside the sloe freezes, it expands and makes the tough skin softer. Therefore, if you're picking the sloes prior to frosty weather, simply pop them into the freezer for a day or so to get the same effect.

Making sloe gin – or vodka – couldn't be simpler. As well as your full bottle of gin (or vodka) you will need a clean, empty bottle.

MAKES 1 X 70CL BOTTLE

450g sloes, washed, dried and frozen
  as above
350g granulated white sugar
70cl bottle gin or vodka
Couple of almonds (optional)

Split the fruit, sugar and gin between the 2 bottles. Seal the bottles and leave in a cool, dark place for about 3 months, shaking the bottles occasionally. If you make the gin in the autumn when the sloes are on the trees, it will be ready by Christmas – if you can wait that long! Some people like to add a couple of almonds, but to be honest

I've tried it both ways and I can't tell what difference they make.

You can either serve the gin with the sloes left in the bottles, which looks very pretty, or decant the alcohol from the fruit and combine the contents of the 2 bottles (there will be a little left over so, of course, it's essential that you test it – I suggest on its own, over crushed ice).

It seems a shame to have to throw away the booze-soaked sloes. So don't! There are a few things you can do with them. You can add them to a bottle of red or white wine and leave for a month to make an unusual fortified wine, for example. But my favourite is to make chocolates with them, as in the following recipe.

Sloe Gin Chocolates

The weight of the gin-soaked sloes given in the ingredients here is prior to the removal of the stones. Once you have cut the stones away, you will end up with approximately half the amount of fruit.

MAKES APPROX. 20 CHOCOLATES

500g dark chocolate (or milk chocolate if you prefer)
500g gin-soaked sloes, stones removed
85g toasted hazelnuts (see page 104) or toasted flaked almonds, smashed up

First, melt the chocolate in a bain-marie or in a heatproof bowl suspended over a pan of simmering water. Then stir in the sloes and the nuts. Take a sheet of greaseproof or parchment paper and put 5cm 'blobs' of the mix onto the paper. Leave to set. *Voilà!*

Easy Sloe Jam

You can also use the gin-soaked sloes in a delicious boozy jam, or you can use fresh sloes, or a mixture of both.

MAKES APPROX. 4 X 454G JARS

2.5kg sloes, before stones have been removed
500g apples (you can use crab apples)
2kg granulated white sugar

Weigh your gin-soaked sloes and, if necessary, top up with fresh sloes that have been softened by soaking overnight and then pressed through a sieve to remove the stones.

Next, wash and grate the apples (including the skins) and cook in a pan with the sloes. Simmer until they're soft and then add the sugar, cooking to reduce the volume by a third. You can either sieve again for an extra-smooth jam, or jar up as it is, using warmed, sterilised jars. Cover with wax discs, and seal.

Apple and sloe is a delicious combination. You can make a sensational bread and butter pudding by spreading the jam on the slices of bread.

Steamy Sloe Gin, Caramelised Apple and Sloe Pudding

Here's another quirky recipe for using sloes and sloe gin. If you haven't made your sloe gin yet, or if you find it's all gone, 'normal' gin will do.

SERVES 4

500g baking apples, peeled, cored and sliced
200g sloes, washed and stalks removed
Caster sugar, to taste, plus 200g to cook with the apples
1 tbsp blackberry jam
175g unsalted butter
150g soft light brown sugar
3 large eggs, beaten
2 tbsp sloe gin
235g self-raising flour
Pinch of salt
Whipped cream or crème fraiche, to serve

Place the 200g caster sugar in a pan with a splash of water and stir over a medium heat and stir until the sugar has just started to brown very slightly. Then add the apples, another splash of hot water, and cook slowly, stirring gently, until the apples are tender and sticky with the caramelised sugar.

Then place the washed, stalk-free sloes in a pan and just cover with water. Bring to the boil and simmer gently on the stove until the sloes break down and become pulpy. Drain the water away and rub the fruit through a sieve into a clean pan. Discard the stones, reheat the sloes and sweeten to taste with caster sugar. Leave to cool.

Butter a 2-litre pudding basin and place a disc of baking parchment in the bottom. Put the apple mixture into the basin, then add 2 heaped tablespoons of the sloes with the blackberry jam. Stir in loosely then smooth up the sides a little.

Cream the butter and brown sugar together until pale and fluffy. Beat in the eggs a little at a time, and then add the gin. Sieve the flour and salt together and fold into the butter and egg mixture. The mixture should be of a dropping consistency. If it's a little dry, mix in a little milk to slacken it.

Spoon on top of the jam in the basin and level off with a spatula. Cover with a disc of buttered baking parchment, then a layer of pleated foil, and either place the basin in a steamer, or put in a large pan with a saucer placed underneath the basin, then fill with sufficient boiling water to come halfway up the sides of the basin. Steam the pudding for 1½–2 hours.

When cooked, turn out onto a serving plate and leave to stand for a few minutes before you remove the mould, making sure that the whole pudding ends up on the plate rather than half of it staying put in the bowl! Serve with whipped cream or crème fraîche.

Sorrel
Rumex Acetosa

*flowers with
central green seed
and crimson edges*

*glossy smooth green leaves,
becoming darker and
slightly leathery with age*

*base of leaves
always pointed
and amplexicaul
to the stem*

*diseased leaves are
common, often flushed
scarlet (they taste as good
as the healthy ones)*

There are several kinds of sorrel, but the one we're looking at here is common sorrel. All the different sorrel family (such as sheep's sorrel and French sorrel) are edible, though.

A perennial plant, sorrel has smooth, slightly shiny leaves that are long and strappy in shape. The leaves vary in size but can grow up to 12cm in length. Long stems bearing noticeable red and yellow flowers appear in the summer, and indeed these dry out to mark the position of the plant during the autumn and winter. Sorrel likes to grow in grassy areas and prefers an acidic soil that's rich in iron.

Culinary uses

Sorrel is a plant that's easily overlooked until you know how useful it is. Chewing sorrel in the summer was once common practice, as Culpeper mentions, since the little, slightly bitter leaves are very good for quenching your thirst. Not many people seem to know this any more, but you should try it. The Romans and Greeks used to nibble the leaves after overindulging, but for a different reason, since the plant has a soothing effect on the stomach.

In fact, sorrel has been used for a very long time as a salad vegetable; there are even records of it as such from the time of Henry VIII. Given the lovely lemony flavour of the fresh young leaves and the abundance of the plants, it's surprising that they're not in more popular use. In general, any foods that taste good with a citrusy kick will sit very happily with sorrel. Try stirring a handful of chopped young leaves into risotto 5 minutes before serving. It's delicious.

When cooking with sorrel, be careful not to use any implements containing iron (this includes stainless steel and non-stick saucepans) since the chemicals in the plant will react with the metal and discolour it. Don't eat *too* much sorrel, though, since the oxalic acid that it contains is mildly poisonous and can interfere with our ability to absorb calcium. Boiling the leaves helps to reduce the amount of oxalic acid.

Medicinal uses

Good old Culpeper, writing in the 17th century, offers the following advice:

'Sorrel is prevalent in all hot diseases, to cool any inflammation and heat of blood in agues pestilential or choleric, or sickness or fainting, arising from heat, and to refresh the overspent spirits with the violence of furious or fiery fits of agues: to quench thirst, and procure an appetite in fainting or decaying stomachs: For it resists the putrefaction of the blood, kills worms, and is a cordial to the heart, which the seed doth more

effectually, being more drying and binding ... Both roots and seeds, as well as the herb, are held powerful to resist the poison of the scorpion ...'

Did you know?

Sorrel is also a name for a boy or a girl. Here, the meaning of the word is 'red-brown' and is derived from the French word, *sorrell.*

Sorrel Soup

SERVES 4

120g butter
1 medium onion, peeled and chopped
600ml vegetable stock (made with a
  stock cube or good-quality bouillon
  powder)
1 large potato, peeled and cubed
450g sorrel
Mixed dried herbs, added to the stock
Salt and ground black pepper, to taste
Single cream and a few chopped fresh
  young sorrel leaves, to garnish

Melt the butter in a large, heavy-bottomed non-iron pan over a medium heat. An enamelled pan is fine for the purpose. Add the onion and sauté until it's soft and golden, making sure it doesn't burn.
Add the stock, bring to the boil and throw in the potato cubes. Simmer for about 15 minutes.

Meanwhile, taking only the freshest sorrel leaves, clean them thoroughly and cut away any tough stems. Chop the leaves, add to the pan and cook for 1 minute. Add some dried herbs and then process in batches in a liquidiser or blender, or use a hand blender. Add salt and pepper to taste. Serve with a swirl of cream, a grinding of black pepper and a sprinkling of finely chopped raw young sorrel leaves.

Sorrel and Goat's Cheese Tart

You can use a ready-prepared pastry case in this recipe but as they're so easy to make, why not spare yourself the expense?

SERVES 4–6

PASTRY CASE
225g plain flour
1 tsp salt
Approx. 115g butter, at room
  temperature
Ground black pepper
Chilled water

Sift the flour and salt into a bowl, then chop in the butter with a knife and rub in with your fingertips, being sure to keep lifting the flour out of the bowl to keep it airy. Add the seasoning and then gradually stir in enough chilled water to make a dough that's not too soggy. Knead briefly, roll into a ball, pop into a plastic bag and put in the fridge for half an hour.

Meanwhile, preheat the oven to 200°C/gas mark 6 and grease a 20cm flan tin.

Dust a work surface with flour and roll the pastry to a thickness of about 3mm. Gently drape the pastry into the flan tin, allowing it to fall into the edges, without stretching it.

Then place the tin on a baking tray without bothering to trim the edges of the pastry (this ensures that there are no shrunken edges later). Prick all over the bottom of the pastry with a fork and bake for about 8 minutes. Remove from the oven, allow it to cool slightly and then trim away the excess pastry.

FILLING
110g goat's cheese (or any strongly
  flavoured creamy cheese)
Good handful of sorrel, coarsely chopped
6–8 shallots, thinly sliced
3 eggs
250ml milk
1 tsp grain mustard
¼ tsp salt
Freshly ground black pepper, to taste
100g strong cheddar, grated

Preheat oven to 230°C/gas mark 8.

Spread the goat's cheese in the bottom of the pastry case. Cover with the chopped sorrel and the shallots.

Beat the eggs, milk and mustard together and season with salt and pepper, then pour over the contents of the pastry case. Sprinkle with the cheddar and then bake for 35–40 minutes or until the top is golden brown. Serve with a salad of wild leaves.

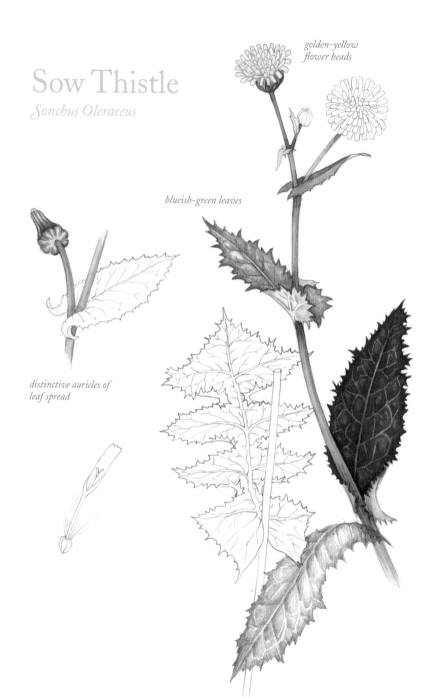

Sow Thistle
Sonchus Oleraceus

golden-yellow
flower heads

blueish-green leaves

distinctive auricles of
leaf spread

Very similar in appearance to the dandelion, the sow thistle even has similar golden-yellow flower heads that appear in May. But the sow thistle can grow much taller than the dandelion, up to 1.5 metres in height. The leaves are spiny, a blueish green on top and a duller greyish green underneath. Sow thistles are not 'real' thistles and are so named because they're good fodder for pigs and other livestock; not only that but the leaves also look a little like pigs' ears.

Culinary uses

Legend has it that the sow thistle was gifted by the goddess Hecate, to Theseus, the king of Athens, who then ate the thistles before doing battle with the great bull of Marathon. Be that as it may, the sow thistle is a tasty and useful vegetable and the roots, like those of the dandelion, can be dried, roasted and ground to make a beverage (follow the method described for dandelion coffee on page 76). The leaves can be chopped and added sparingly to salads (the raw texture is a bit dry) or cooked by steaming, boiling or sautéeing and used as any other green vegetable.

A word of warning: the stems of sow thistle contain a milky sap, very like that of dandelions which can be a skin irritant. Be careful that the sap doesn't make its way into your recipes because it doesn't taste very nice.

Medicinal uses

Parts of the sow thistle have been used as a diuretic, to encourage menstrual flow and even in combating cancer.

Did you know?

The *sonchus* part of the Latin name means 'hollow', and refers to the hollow stem of the plant.

Sow Thistle Cannelloni

SERVES 2 AS A MAIN COURSE OR
4 AS A STARTER

Good glug of olive oil
1 large onion, chopped
115g mushrooms, chopped
7 medium garlic cloves, crushed
A good pinch of dried basil and oregano
1 can chopped tomatoes
Good couple of handfuls of sow thistle
* leaves, washed and dried*
8 ready-made cannelloni shells
425g ricotta cheese
425g grated mozzarella cheese
Salt and pepper, to taste

Preheat oven to 220°C/gas mark 7.

Heat the oil in a frying pan and
sauté the onion until it's transparent.
Add the mushrooms and fry for
another couple of minutes. Add
the garlic and herbs, a little of the
tin of tomatoes, and season with salt
and pepper. Add the sow thistle
leaves and a splash of water and cook
for another couple of minutes until
the tomatoes begin to thicken. Allow
the mixture to cool and then stuff
into the cannelloni shells.

Put the stuffed shells neatly into an
ovenproof dish, then spoon the rest
of the tomatoes over the top. Mix
the cheeses together and spread over
the top.

Cover and bake for half an hour.
Remove the covering and cook for
another 10 minutes, or until the
cheese is golden and bubbling.

Sow Thistle Spread

MAKES APPROX. 150G

Good glug of olive oil
2 onions, chopped
250g mushrooms, chopped
2 handfuls sow thistle leaves, washed
  and dried
Salt and pepper, to taste
Soy sauce, to taste

Heat the oil in a frying pan and sauté the onions until soft and translucent. Add the mushrooms, turn up the heat and cook for another minute. Throw in the leaves and the salt and pepper and a splash of soy sauce, and cook fast for another couple of minutes. Leave to cool, then blend with a hand blender until smooth, adding more soy sauce and seasoning to taste.

This spread is great on toast, on crusty bread or on top of jacket potatoes.

Violet
Viola Odorata

*distinctive heart-shaped leaves
start bright green, becoming a
dark blue-green with age*

Violet flowers are not always violet in colour. They can be pink, dark indigo blue or even white, and there's even a very rare strain that is blood red. The flowers start to appear in late March and early April. The violet has distinctive heart-shaped leaves with softly scooped edges.

Culinary uses

There are several kinds of violet in the UK and you can use any of the flowers decoratively, but the only violet with that distinctive scent and taste is the sweet Violet, or *viola odorata*. You can identify this type not only by its leaves, which get larger after the plant has flowered, but by that unmistakeable scent.

Only use sweet violets if you are sure that you can gather them without harming the rootstock of the plant. They are delicious, it's true, but look so much better in the hedgerow than on a plate, in any form.

As you might imagine with such a magically beautiful wild flower, there are many myths and legends attached to it. The great god Jupiter is said to have invented the violet as food for his adored lover, Io, to protect her from the attention of his jealous wife, Juno. 'Io' gave her name to the violet and also to one of the chemicals that causes its gorgeous scent, Ionone. The Romans believed that violets prevented drunkenness, so they used the flowers to make a sort of homeopathic-remedy wine, no doubt in the hope that they would be able to get merry without suffering any consequences.

Indeed, the making of violet wine, tea and syrup was once very popular but seems to have dwindled in more recent times. This might be because there are fewer violets around than before.

Violet leaves have a mild 'green' flavour and can be added to salads.

Medicinal uses

Violets are believed to alleviate depression, remove headaches and cure insomnia. A tisane made of the leaves is used as a remedy for bronchitis, catarrh and even rheumatism; if you want to try this, simply infuse a handful of the leaves in hot water, strain and add honey to taste. And, even more interestingly, the 1983 edition of the British Pharmacopoeia records the use of sweet violet, taken both internally and externally, as a way of treating cancer.

Did you know?

The Victorians, who had an elaborate language of flowers, would send violets as a message of modesty and faithfulness.

Crystallised Violets

Crystallised violets make a beautiful decoration for cakes and sweets and are very easy to make. Whisk an egg white with a little rose water or lavender water, and then use to coat the flowers with a paintbrush. Make sure you leave the stems on the flowers since this makes them a lot easier to handle. Dip the flowers into caster sugar and leave overnight on blotting paper to dry. It's also nice to give the leaves the same treatment.

Violet Sugar

½ jar violets
½ jar granulated white sugar or
  caster sugar

Simply layer the violets and the sugar in a jar and leave in a dark place for a week, then pour through a sieve to remove the flowers. The sugar will be impregnated with the flavour of the flowers.

Violet Cream Syllabub

SERVES 4–6

4 tbsp rice flour
2 tbsp ground almonds
1 litre full-fat milk
4 tbsp granulated white sugar
Zest of 2 oranges
Handful of sweet violet flowers
Handful of toasted pistachios
  (not salted ones)
A little whipped cream,
  for topping
Crystallised sweet violets,
  to garnish

Mix together the rice flour and almonds with half the milk. Bring the remaining milk to the boil in a saucepan and add the rice flour and almond mixture. Simmer for 10 minutes, stirring occasionally and making sure that it doesn't burn. Add the sugar and most of the orange zest (set aside a little for decoration) and then simmer the mixture until it's the texture of custard. Let it cool.

While the 'custard' is cooling, tear the flowers roughly into quarters (don't chop, since blades can discolour the flowers) and stir them into the mix. Whisk lightly, pour into champagne glasses and leave the mixture for a couple of hours to absorb the flavour of the flowers.

Before serving, sprinkle the chopped nuts and reserved orange zest on top of the pudding. Add a dollop of whipped cream and perch the crystallised flowers on top of the cream – add the leaves too if you like.

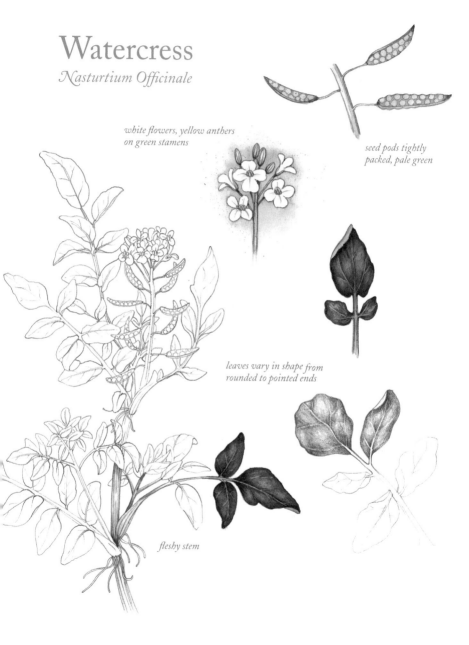

Watercress
Nasturtium Officinale

white flowers, yellow anthers
on green stamens

seed pods tightly
packed, pale green

leaves vary in shape from
rounded to pointed ends

fleshy stem

Dense, mat-like drifts of watercress like to inhabit the marshy ground at the edges of fast-flowing rivers and streams. The leaves are oval with crinkly, wavy edges, graded in size symmetrically along stalks that curve outwards. Watercress proliferates from March to October (except where it might be damaged by frost), and flowers from May onwards to the end of its season. Those flowers are small and white, each with 4 petals, and form a cluster at the top ends of the plants. Watercress doesn't seem to grow at any significant altitude.

Many of the places that were once famed for their watercress crop can be identified by name – if you know what to look for. These places include anywhere with the suffix 'kers' or 'kess', meaning 'cress'; such places include Kersal in Lancashire and Kersey in Suffolk. Once the plant started to be cultivated in the 19th century (the first place to do so was Gravesend), special train lines were constructed to carry the crops to London – the so-called Watercress Lines. It's hard to imagine, now, just what an impact this plant must have had, or how esteemed it was, that it needed its own special mode of transport.

Culinary uses

When gathering cress, be aware that if there are cattle and sheep nearby, the cress could be contaminated by liver fluke. If you are going to eat any wild vegetable raw it will need to be washed, but be especially careful with watercress. Cooking, however, destroys the fluke. Otherwise, be sure to cut the plants rather than pulling them so that you don't damage the rootstock. The more mature shoots are the better bet for cooking.

Cultivated watercress hardly varies from the wild form. It has a deliciously fresh, distinctive taste, peppery and sharp.

Medicinal uses

Culpeper recommended watercress for cleansing the blood and for stimulating the appetite. It's packed with vitamins A and C, folic acid, iron and calcium and is low in fat. The plant is also used as a remedy for dermatitis and eczema, applied externally. Incidentally, when using watercress, be sure to tear the leaves, since cutting them with a blade can destroy some of the vitamin C content.

Did you know?

Recent research from Southampton University found that eating watercress increased the number of cancer-fighting molecules in the blood.

Cheesy Watercress Scones

MAKES APPROX. 12 SMALL SCONES

100g self-raising wholemeal flour
100g self-raising plain flour
½ tsp salt
100g butter, at room temperature, cubed
50g strong cheese, grated
85g watercress, torn up

Preheat oven to 200°C/gas mark 6. Line a baking tray with baking parchment.

Sift the flours and salt together and rub in the butter. Stir in the cheese and the torn watercress and add a little chilled water to make a soft dough. Roll out on a floured work surface to about 2.5cm thick. Cut into 5cm circles, using a pastry cutter.

Place on the baking tray and bake for about 15 minutes, until risen and browned. Serve the scones hot and dripping with butter.

Watercress Hummus

MAKES APPROX. 800G

2 x 410g cans chickpeas
1 garlic clove, crushed
3 tbsp olive oil
1 tbsp tahini
Juice of 1 lemon
Good pinch of cayenne pepper
85g watercress, torn up
Salt and black pepper, to taste
Pitta bread or crudités, to serve

Drain the chickpeas, reserving 100ml of the liquid. Put half the chickpeas and the 100ml liquid in a food processor, add the garlic, olive oil, tahini, lemon juice, cayenne pepper and plenty of salt and freshly ground black pepper. Whizz until the mixture is smooth.

Add the remaining chickpeas and the watercress and whizz again, but not too much this time, as you want to leave the hummus quite rough and chunky. Serve with warm pitta bread or as a dip with crudités.

Watercress, Melon and Strawberry Smoothie

SERVES 4–6

1 ripe galia melon, cut into chunks
85g watercress, torn up
200g strawberries
Sparkling water or thin plain live
  yoghurt, to taste

Whizz the melon, watercress and strawberries in a blender, and add sparkling water or, if you prefer, the yoghurt at the blending stage instead. Health in a glass!

Watercress Virgin Mary

SERVES 4

100g watercress, torn up
150ml freshly squeezed orange juice
400ml good-quality tomato juice
Tabasco sauce, to taste

Place the watercress and half the orange juice in a blender and blend until smooth. Add the remaining juice and whizz again. Season with Tabasco sauce to taste.

Spicy Watercress and Coconut Soup

SERVES 4

Large knob of butter
Good glug of sunflower oil
2 garlic cloves, chopped
1 large red chilli pepper, de-seeded and chopped
Knob of ginger, grated
1 onion, chopped
1 large potato, peeled and cubed
600ml boiling water
1 sachet creamed coconut, broken up
200g watercress, torn up
Salt and pepper, to taste
Plain yoghurt, to serve

Heat the butter and oil in a heavy-bottomed pan, then add the garlic, chilli and ginger. Heat until the ingredients are popping, then throw in the onion, turn the heat down and sauté until the onion is soft. Add the potato and the water and cook for 20 minutes or until the liquid has reduced by about a quarter. Once reduced, stir in the coconut to thicken it further, then add the watercress. Cook on a low heat for another 10 minutes, then whizz with a hand blender.

Season to taste and serve with a swirl of plain yoghurt.

Watercress and Walnut Pesto

MAKES 120G

30g watercress leaves
1–2 garlic cloves, crushed
40g freshly grated pecorino cheese
30g blanched walnuts
75ml extra-virgin olive oil

Simply blitz all the ingredients in a blender. Delicious served with fresh tagliatelle or stirred into quinoa, or slathered onto hot home-made bread straight from the oven.

Wild Garlic (Ramsons)

Allium Ursinum

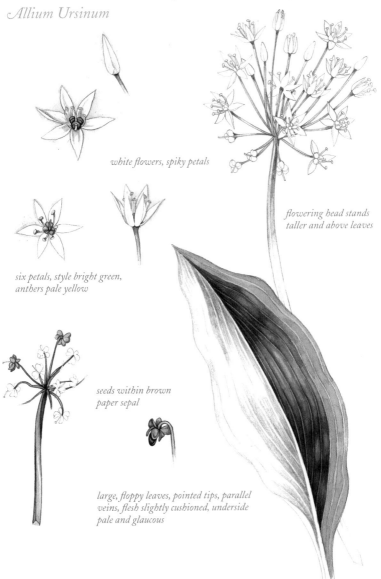

white flowers, spiky petals

*six petals, style bright green,
anthers pale yellow*

*flowering head stands
taller and above leaves*

*seeds within brown
paper sepal*

*large, floppy leaves, pointed tips, parallel
veins, flesh slightly cushioned, underside
pale and glaucous*

Rather than asking what this plant looks like, you might be better to consider the smell, because you're likely to encounter this before you see anything at all. The folk names of the plant include 'stink bombs' and 'stinking nanny' for very good reason! The leaves are about 25cm long, soft, smooth and shiny and tapering to a point. The white starry flowers stud these carpets of leaves from April onwards. The flowers, though, herald the end of the garlic season; shortly after they appear the leaves start to wither and die back, the flowers mature and droop away. You often see wild garlic growing in the same sorts of places as bluebells; the edges of shady woods or out in the open amongst thinner groupings of trees or on the banked verges of hedgerows. By the way, you'll often see wild garlic referred to as 'ramsons'.

Culinary uses

Although the smell is very pungent in the open, the flavour of wild garlic isn't as powerful as the 'normal' variety, but nevertheless a good handful will still give an amazing flavour to just about anything you care to put it in. The bulbs are small, but edible; however, it's the leaves of the plant that are mainly used. The pretty white flowers are also edible and make a good garnish for soups, etc.

Medicinal uses

Wild garlic has all the same medicinal benefits as cultivated garlic, (i.e. it's an insecticide, has antioxidant properties, and boosts the immune system) with the addition of the wild 'magic' that belongs to anything that you have found, identified and used for yourself. It's also very good for the digestion; one of the names of the plant is 'bear's garlic' since bears used to eat it after their winter hibernation to kick their insides into gear!

Did you know?

Ramsons, the common name for wild garlic, leaves a legacy in the names of the places where it once grew prolifically. These include Ramsey on the Isle of Wight, Ramsgate in Kent, and Ramsdell in Hampshire. Turn to any page in an atlas of the UK, and chances are you'll find a 'ram' place name. This is because the presence of wild garlic is a key indicator of ancient woods, and the garlic plants that you pick to use in the recipes here could well be the offspring of ancestors over a thousand years old.

Garlic Flu Remedy

*12 whole wild garlic plants, bulbs
   included, washed well*
600ml water
Zest and juice of 1 lemon
Good dollop of honey

Boil the garlic in the water with
the lemon zest until the water has
reduced by half. Allow to cool a
little, blend without straining, then
add the lemon juice and honey.
Drink a cupful 4 times a day.

Garlic-Infused Oil

This recipe beats all those fancy
gourmet flavoured oils into a pulp,
and costs pennies in comparison.
Simply take a handful of the young,
strappy garlic leaves, wash and dry
thoroughly. Then roll them up like
cigars and pop into the top of a bottle of good-quality extra-virgin olive
oil. Leave for a couple of weeks to
allow the flavour to permeate, before
using. You can pour it into bowls,
add a splash of balsamic vinegar and
some garlic flowers and use as a dip
for crusty home-made bread.

Wild Garlic and Potato Soup

SERVES 4–6

25g butter
Couple of handfuls wild garlic leaves
2 medium-sized potatoes
800ml vegetable stock
Salt and pepper, to taste
Fresh cream, to serve
A few garlic flowers, to garnish

Melt the butter in a heavy-bottomed pan. Tear the garlic leaves into strips, throw them into the pan and cover with a lid. While the garlic is becoming tender, peel and dice the potatoes. When the garlic leaves are wilted, add the potatoes and the stock. Bring to the boil, then simmer for 20 minutes, or until the potatoes are really soft. Season with salt and pepper, then blend with a hand blender. Allow to cool ever so slightly, then ladle into dishes, adding a swirl of cream and a sprinkling of garlic flowers.

Wild Garlic Pesto

MAKES APPROX. 500G

200g wild garlic leaves
200g leaves any other fresh herb
   (watercress or sorrel would be good)
A little olive oil
200g very finely grated hard cheese
200g pine nuts, coarsely ground
Dash of sea salt

Wash and dry the leaves thoroughly, then tear up and put in a blender. Cover with olive oil (not too much – you can always add more if you need it) and whizz until smooth. Stir in the cheese and the pine nuts and add a dash of sea salt.

Ramsons and Goat's Cheese Hors d'Oeuvres

SERVES 4–8 AS NIBBLES

Good handful of wild garlic leaves,
* washed and dried*
350g soft goat's cheese
300g sesame seeds, lightly toasted

Chop the garlic leaves very finely – use a blender if necessary. Mash them into the goat's cheese and then shape into small balls the size of marbles.

Roll the balls in the sesame seeds until they are thoroughly covered, then place in the fridge to chill. Remove from the fridge half an hour before serving.

Wild Garlic Bannock Bread

MAKES ONE SMALL LOAF OR
6 SMALL ROLLS

This recipe, and the wild garlic and greens soup that follows, come from Sarah Howcroft, a shaman and bushcraft teacher based near Brecon in Wales. Served together, the bread and the soup are just scrummy, and the added benefit is that both can be made over a campfire!

Bannock is said to have originated from the Celts. It was favoured by nomadic tribes because the dry mix stayed fresh for long periods, and the fat or oil, water and leaves would be added at cooking time, along with any other flavourings that might have been to hand. All the dry ingredients can be pre-mixed and transported in a zip lock bag, ready to just add water and the ramsons, mix in the bag, and cook on the camp fire.

225g plain flour
110g milk powder
Good pinch of salt
284ml water
Large handful of wild garlic leaves,
* torn up*
1 egg and/or 2 tbsp oil or butter
* (optional)*

Mix all ingredients together into a dough, then shape into a loaf or rolls (which will cook slightly faster) and bake – this is best baked in a Dutch oven over the embers of the camp-fire, but some cooks prefer to fry their bannock dough in a frying pan (cast iron is best), while others deep-fry it. You can also drop spoonfuls of dough into a stew to produce something resembling dumplings. If you want to get really Wild West, take a stick from a (non-toxic) tree, twirl the dough over the end of it and use the stick to cook the bread in the fire. Scrumptious, and children *love* making this bread out in the open.

Wild Garlic and Greens Soup

You can use any wild greens that you can find – nettles, sorrel, cleavers, garlic, dandelions – depending on the time of year. Young, tender leaves are best, as older ones can be stringy. You can also add green vegetables such as leeks.

The size of the pan and the abundance of leaves, and the appetite of the diners will dictate how many people this soup will serve.

Large saucepan of wild greens
2 big handfuls wild garlic leaves
Knob of butter
1 onion, finely diced
1 potato, peeled and finely diced
Swiss vegetable bouillon powder to taste
  – at least 3 heaped tsp
Black pepper or other seasoning, to taste

Roughly tear the greens and the wild garlic.

Melt the butter in a large heavy-bottomed pan and sweat the onion. Add all the other ingredients, cover with a little water, and cook for about 10 minutes. Don't add too much water to begin with, as the greens will cook down and you can then thin the soup according to taste. Blend it with a hand blender for a smoother finish. Finish the soup with a swirl of olive oil, yoghurt or sour cream, or grated cheese.

Wild Strawberry
Fragaria Vesca

*bright-red,
jewel-like fruits
drooping on long
slender stalk*

*small, milky-white
five-petalled flower*

*deeply serrated,
symmetrical leaves*

The wild strawberry looks like a doll's house version of its cultivated cousin. Only 1/2–2cm long, wild strawberries will sprawl across dry, slightly dusty ground, on verges and at the edges of tracks, and sometimes on steep banks. The leaves, too, are a perfect miniature replica of a regular strawberry leaf. The pretty star-like white flowers have 5 petals and appear in early summer, followed by their glistening, jewel-like fruits.

Culinary uses
Nothing can compare with the taste of wild strawberries. Sometimes, bizarrely, they taste as though they already have cream on them!

Wild strawberries are so tiny, and so beautiful, that it seems criminal to reduce them into pulp by cooking them. Therefore the only recipe I have included for these little gems is inspired by the renowned jelly-mongers, Sam Bompas and Harry Parr, and their book *Jelly with Bompas and Parr*, published by Anova Books, 2010. This recipe keeps the strawberries intact.

Since the strawberries are free, spend the money you've saved on the best champagne you can afford to make this luxurious piece of edible art which really is a 'special occasion' recipe … but the brief appearance of wild strawberries alone should be something to celebrate.

In the recipe opposite, a bottle of champagne will make enough jelly for 4 proper servings – and a glass for yourself while you are cooking. Make sure you chill the champagne ahead of time – it's important that you drink on the job, according to Sam and Harry.

Medicinal uses
Strawberry leaf tea can be used as a remedy for diarrhoea, but beware; some people are allergic to these leaves. This tea makes a good substitute for ordinary black tea, and the leaves can be dried or used fresh. The fruit contains vitamin C.

Did you know?
You might assume that the name of the strawberry comes from the straw that's placed around the base of the plants to protect them from slugs and snails, but it actually has very different origins. It comes from the Anglo-Saxon 'streow berrie'. Here, 'streow' means 'stray' and describes the meandering growth of the plant as it sets its runners in every direction.

Champagne and Wild Strawberry Jelly

MAKES 500ML

*5 sheets leaf gelatine or equivalent
  amount of agar-agar/vegetarian
  gelling agent
450ml champagne or sparkling wine
40g caster sugar
Squeeze of lemon juice
Good handful of wild strawberries, plus
  extra to serve*

Cut up the leaf gelatine and put it into a heatproof bowl that's big enough to take the entire quantity of champagne. Cover the gelatine with champagne taken from the bottle – about 100ml should do it. Leave for ten minutes so that the gelatine goes soft.

Bring a pan of water to the boil and put the bowl of softened gelatine on top of the pan of boiling water so that the gelatine melts completely. Once this has happened, stir in the sugar until it has dissolved and add another 100ml champagne to the champers/gelatine mixture.

Combine the melted gelatine/champagne/sugar mix with the rest of the champagne by pouring it through a sieve to remove any un-melted lumps, and then pour into a measuring jug. Squeeze the lemon through the sieve, too. Top up with champagne until you have 500ml.

Then place the washed and dried strawberries in the cocktail glasses, and pour in the champagne. Allow to cool and then put in the fridge.

Before serving, decorate with a few extra wild strawberries. You could also crystallise the leaves (see method used for violet flowers on page 196) and even dust them with gold leaf.

Wimberry

Vaccinium Myrtillus

*flower bright
scarlet, drooping*

unripe berry

(Also called bilberry, blaeberry, whortleberry and sometimes huckleberry)

If you see heather growing, then it's not unlikely that the habitat will be right for wimberries, too. This is the fruit of the truly wild environment. Wimberries are low-growing, shrubby plants, with upright stems that occasionally reach up to 60cm high. The leaves are bright green, circular, with tiny teeth, and on occasions can measure up to 1cm long. The flowers that later give way to the bright shiny black/purple/indigo berries are pinky-green and bell-shaped. The fruiting season is generous, from July to September and sometimes on into October.

Culinary uses

Harvesting these delicious, flavoursome berries was once an important occupation in the rural areas where the berries grow, and the whole family would be united in this endeavour, the children sometimes taking days off school just to pick the crop. These days, those in the know, and the commercial pickers that supply restaurants and shops in their local area, save time by using a many-pronged fork, but this enterprise seems to be dying out. Even if you find a particularly abundant patch, picking bilberries by hand is slow going, but worth it.

I'm lucky enough to live in wimberry land. I remember on one occasion I got up very early to pick the berries and after a couple of hours I had collected 2 large tubs, full to the brim. I edged further and further away from these tubs only to find on my return that my dog had eaten the entire contents. They're so tasty, I couldn't really blame him.

You'll see that the Latin name for this plant contains the word *myrtillus*. In France the fruit is called *myrtilles*, and if you're over there you may find jam made from it.

Medicinal uses

Berries really are a superfood, containing large amounts of vitamin C. But the wimberry has an additional benefit. It contains anthocyanins; the word 'cyan' should give a clue since it's contained in the blue pigmentation of the skin of the berry. It's an antioxidant, which protects against UV rays. The fruits are also an effective cure for diarrhoea – just nibble a handful if you're suffering – while the leaves have been used in the treatment of diabetes.

Did you know?
In his book *The Return of the Native*, published in 1878, Thomas Hardy refers to wimberries as 'blackhearts'. In Ireland, where they're known as *fraughan,* they are traditionally collected on the last Sunday in July, Fraughan Sunday.

Wimberry Soup

(Also known as 'blaebarsuppa' from the common name for the berry in Scandinavia, which translates as 'blueberry'.)

In Scandinavia, wimberries enjoy almost cult status, and the Scandinavians make a sensational soup of them. A sweet fruit soup might sound a strange concept, but wait till you try it … it's scrummy.

SERVES 4

1.75 litres water
250g wimberries
Approx. 150g granulated white sugar,
*   plus extra for serving*
30g cornflour
Fresh cream, for serving

In a large, heavy-bottomed saucepan, bring the water to the boil and put the berries in it. Reduce the heat and cook until the berries burst – this will take only a couple of minutes. Add the 150g sugar and stir until it's melted.

Mix the cornflour with a splash of cold water to form a smooth paste. Prime the paste with some of the soup, mixing together well, and then add this to the soup. Increase the heat a little, stirring constantly until the soup has thickened.

Pour into bowls and sprinkle a little extra sugar on top. Serve either warm or cold, with a swirl of cream. If serving cold, a few shredded mint leaves (wild water mint, naturally) makes a great garnish.

Wimberry Vodka

This is sensational. But make sure you use a good-quality vodka in the first place. You can use gin instead, if you prefer. As well as a full bottle of vodka, you'll also need a clean, empty bottle.

MAKES 1 X 70CL BOTTLE

70cl bottle vodka
Enough wimberries to fill a whole vodka bottle
200g granulated white sugar (if you want it – it's not really necessary)

Pour half the vodka into the empty bottle, top up both bottles with wimberries. If you want to add the sugar, do this before adding the berries.

Then leave for a couple of months, shaking the bottles occasionally. You can either serve with the berries in the bottle, or strain the vodka to make just over 1 full bottle – there will be enough left over for you to test it. Great to make for Christmas!

Wimberry Ice Cream

The colour of this ice cream is extraordinary. It's a proper mid-purple and looks quite surreal.

SERVES 4

Good handful of wimberries, plus extra
*   to garnish*
290ml whipping cream
290ml milk
2 egg yolks
100g caster sugar
Dollop of crème fraîche
Wild mint leaves, to garnish

Wash and dry the wimberries and whizz in a blender. Then whisk the cream separately.

Boil the milk in a saucepan, take it off the heat and whisk in the egg yolks, fold in the cream sugar, wimberries and crème fraîche.

You can then throw the whole caboodle into an ice-cream maker and follow the manufacturer's instructions, or you can put in a tub in the freezer and give a good stir every half an hour, for a 2–3 hour period, to prevent crystallisation.

Serve with a garnish of a few extra wimberries and some mint leaves.

Wimberry Trumble

This recipe comes courtesy of Joy and Doug, proprietors of The Traveller's Rest in Talybont-on-Usk. Usually made by June, it's a sensational cross between a tart and a crumble!

SERVES 6

PASTRY CASE
225g plain flour (or, better still, a 50/50
*   mix of flour and finely-ground pecan*
*   nuts)*
50g white caster sugar
110g butter, at room temperature
Chilled water

FILLING
170g fresh wimberries
50g granulated white sugar

TOPPING
110g plain flour
50g caster sugar, plus a little extra
A little ground cinnamon (if you fancy
*   it)*
50g butter

To make the pastry case, put the flour (or flour/pecan nut mix) and the sugar into a large bowl. Mix well and then rub in the butter with the tips of your fingers, remembering to lift the flour out of the bowl to make a light and airy pastry. Add the chilled water a little at a time, until you have a dough. Knead for a minute or so, then form into a ball and put in a plastic bag. Refrigerate for half an hour.

In the meantime, preheat the oven to 170°C/gas mark 3 and grease a 20cm pie dish with butter.

When the pastry is chilled, roll it out on a floured surface to about 3mm thick and carefully drape it in the pie dish, letting it fall into the edges of the dish. Prick the bottom with a fork and bake blind for 5 minutes.

For the filling, mix the wimberries with the sugar and spoon into the pastry case.

Mix together the ingredients for the topping, rubbing in the butter to make coarse crumbs. Spread it evenly over the fruit, then sprinkle a little extra sugar over the top.

Bake for 20 minutes, or until the top of the trumble is golden and the fruit is starting to ooze through it a little. Leave to cool, and serve with custard, ice cream, whipped cream – or a little bit of all three!

Fraughan with Athol Brose

Here, I've combined the wimberries – known in Ireland as *fraughan* – with an ancient Scottish recipe, 'Athol Brose', which is sometimes referred to as 'crannachan'. You can replace the whisky with rum or sherry if you prefer.

SERVES 4–6

100g pinhead oatmeal
4 tbsp whisky
4 tbsp clear heather honey
500ml double cream
350g wimberries

Steep the oatmeal in the whisky overnight. The result should be a firm-ish paste. Then stir in the honey and a little more whisky, if necessary. Leave for a couple of hours, and then stir in the cream.

Layer the cream mixture and the wimberries in cocktail glasses, and add a further dash of whisky if you fancy it.

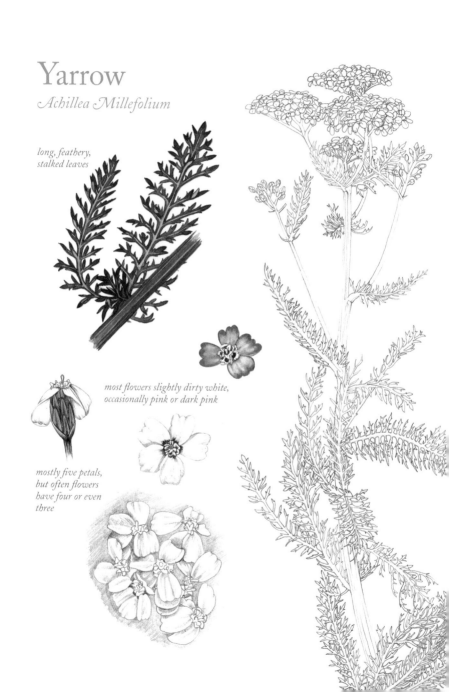

Yarrow

Achillea Millefolium

*long, feathery,
stalked leaves*

*most flowers slightly dirty white,
occasionally pink or dark pink*

*mostly five petals,
but often flowers
have four or even
three*

A rather pretty if often overlooked wild flower, the yarrow can be any size from 8–40cm tall. The leaves are delicately feathery, rather like a butch asparagus fern, and the plant forms clumps that can sometimes be quite large. The stalk that bears the flower is long and straight, and the many tiny flowers (hence the *mille-folium* part of the Latin name) are borne in thick flat-topped clusters at the tops of these stems. The flowers can be white or pale-ish to a darker pink. Yarrow has a pungent, aromatic smell that is enhanced by lightly crushing the leaves between your fingers.

Yarrow grows all over the UK, in lawns, fields, hedgerows and on verges.

Yarrow stems have long been used by practitioners of the ancient divinatory arts of the I Ching, cast on the ground by the questioner and analysed by the diviner. Also, bundles of the plant were hung over doorways to avert illness and/or bad luck in the household.

Culinary uses

Yarrow leaves have a dill-like aromatic quality and can therefore be used as an interesting alternative. They are good in mixed salads; a little lemon juice and sugar really helps to bring out the flavour.

A tea made of yarrow, with just a couple of leaves per mug, a spoonful of sugar and boiling water, is not only tasty, but also effective against colds.

Yarrow is easy to dry, by suspending the stalks upside down (this helps the volatile oils flow down to the tips of the plant) in a paper bag. Once they are dry, crumble the leaves and store in a dark glass jar, or a transparent jar with brown paper wrapped around it.

Medicinal uses

The *Achillea* part of the Latin name comes from Achilles, the great Greek warrior renowned for his one weak spot, namely, his ankle. Yarrow was used by this great fighter to heal the wounds of his men during battles; and indeed the yarrow has a deserved reputation as a healing plant, since it contains anti-inflammatory chemicals. Another clue is found in its common names, 'staunchweed' and 'soldier's wound wort'. Yarrow is also used to boost the circulation. And if you have toothache, simply bite on some fresh yarrow leaves and hold them on the tender place; it really does help.

Did you know?

Yarrow has insecticidal qualities, which makes it a good companion plant. You can also put a bundle of it inside your wardrobe or clothes drawer to keep creepy crawlies away.

Yarrow Beer

This isn't officially a 'beer', more of a fizzy, slightly alcoholic lemonade made in a similar way to ginger beer. But it's lovely to drink on a hot summer's day. You'll need a large food-grade container, demijohns and airlocks, and a siphon.

180g dried yarrow
Lump of ginger, thumped
Juice and rind of 1 unwaxed lemon
5 litres bottled or filtered water
450g demerara sugar
450g jar honey (not the commercial stuff that's been boiled – buy some from a local maker at a country fair or farmers' market)
25g cream of tartar
2 tsp yeast nutrient
7g sachet wine yeast, made up to manufacturers' instructions

Put the yarrow, bruised ginger and lemon rind into a large pan with the water. Bring to the boil, simmer for 10 minutes, adding the sugar and the honey in small batches and letting it dissolve. Add the cream of tartar. Remove from the heat and allow the liquid to cool to blood temperature, then stir in the yeast nutrient and strain the liquid from the solids.

Pour the liquid into a large food-grade container. Add the lemon juice, and then, when the liquid has cooled a little (so it's not too hot to stick your finger in), add the yeast. Cover and leave overnight in a warm, dry place.

Siphon into demijohns with airlocks and leave for a further week, then decant into bottles and chill. The drink doesn't last long, so enjoy it in the sunshine.

Roast Mediterranean Vegetable Couscous with Yarrow

SERVES 4

Good glug each of vegetable oil and olive oil
2 large onions
1 head garlic
1 red pepper
1 yellow pepper
1 orange pepper
1 courgette
500g cherry tomatoes, left whole
Glug of balsamic vinegar
500g couscous
Knob of butter
Handful of yarrow leaves, plus extra to garnish
Juice of 1 lemon
Freshly ground black pepper, to taste

Preheat oven to 200°C/gas mark 6. When hot, pour a good glug of each oil into the bottom of a large, square roasting tin and place in the oven until the oil is hot. Add the onion, chopped into rough wedges. In the meantime, slice the bottom from the garlic, separate the cloves, but leave the 'paper' on, wrap in foil and pop into the oven, checking from time to time.

Stir the onions a couple of times to stop them burning, then add the peppers and courgette, sliced into chunks of roughly the same size.

Add the tomatoes, then stir in the balsamic vinegar and turn the oven down a notch. Cook for half an hour, or until roasted, turning the vegetables occasionally.

Once the garlic cloves are tender, take them out of the oven and leave to cool slightly, then remove the skin and add the cloves to the cooked vegetables.

Meanwhile, steam the couscous, add the butter and stir in the raw yarrow leaves. Arrange the couscous on a large platter, with the vegetables sitting on top, garnished with more yarrow leaves. Add the lemon juice and plenty of black pepper, and serve.

Other Ideas

Here are a few more suggestions for
using flowers, leaves and berries from
the hedgerow.

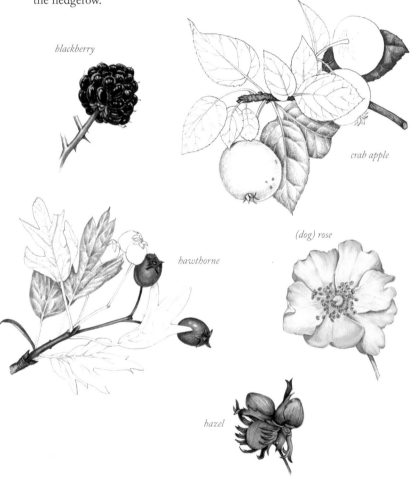

blackberry

crab apple

hawthorne

(dog) rose

hazel

Wild Flower Ice Bowl

Why not serve up salads or desserts made from hedgerow ingredients in a beautiful bowl, made of ice, with wild flowers and leaves suspended in it? It's remarkably easy as long as you have a freezer with a good-sized drawer in it.

You'll need 2 bowls that stack inside one another, water, something heavy like a rock or a couple of cans of beans, and various fresh leaves, flowers and even twigs.

Pour water about 5cm deep into the larger bowl, and freeze. Then rest the smaller bowl on top of the ice, weighted down with something heavy. Put your leaves and flowers into the space between the 2 bowls, and half-fill with water. Freeze again. Then repeat (so you have an even distribution of the flowers and leaves through the ice, otherwise they all tend to float to the top). Freeze again. When the water is solid, run hot water into the top bowl to help it lift from the ice, then turn the bowl upside down and carefully run water over the bottom of the first bowl, holding onto the icy part with your other hand, since it will drop free and you don't want it to smash.

These ice bowls not only look beautiful, adding a real 'wow' factor to any party, but they last a long time and can be used again. I have even served flaming Christmas pudding in one with no disasters.

Again on the subject of ice, keep a ready supply of frozen wild summer fruits in your freezer to add to spirits (vodka, gin, white rum, etc.) at Christmas, instead of ice cubes.

Brecon Beacons Hedgerow Jam

This is a fabulous recipe for a 'jam' made using many of the ingredients found in this book. It's so easy and it tastes sensational.

It's possible that not all these fruits will be ripe at the same time, so collect and freeze until you have everything you need.

MAKES 10–12 454G JARS

450g crab apples, cored and chopped
250g rose hips
250g hawthorn berries
250g sloes
250g rowan berries
450g elderberries
450g blackberries
125g hazelnuts or beechnuts, finely
  chopped
Granulated white sugar (you'll need at
  least 900g – see below for quantity)
Good knob of butter

Wash the fruit well. Put all the fruit, except for the elderberries and blackberries, in a large pan and add just enough water to cover. Simmer the fruit until it's soft – this will take up to half an hour. When it's soft, push the fruit pulp through a sieve to get rid of any seeds or pips, then weigh the fruit.

Put the fruit back in the pan and add the elderberries, blackberries and the nuts. Bring to a simmer and cook on a lowish heat for 15 minutes. Add 900g sugar plus extra sugar that's equivalent to the weight of the original fruit pulp. Stir, and cook again over a low heat until the sugar has dissolved. Add the butter and bring to the boil. Let the mixture boil rapidly for about 10 minutes and then test to see if it's ready: just put a blob of jam on a chilled saucer and push it with your finger; if the surface wrinkles, then the jam is done. If not, boil for a little longer.

If there's any scum on the top of the jam, remove it, and then spoon the jam into warmed, sterilised jars, cover with wax discs, and seal. Store in a cool place.

Fruit Butters

Fruit butters are a bit like jam, only much easier to make and with a smoother texture. You can use different kinds of fruits or combinations thereof, and so hedgerow fruits lend themselves well to this recipe. Try flavouring with flower syrups, too, or adding spices.

Make sure the fruits you use are wholesome and sound, with no rotten or overly mushy bits. Prepare the fruits as follows:

Crab apples – peel, core and quarter. Too bitter to use on their own, crab apples are good combined with sweeter fruits.
Damsons – chop in half and remove the stones.
Sloes – soak in hot water, then sieve to remove the stones.
Raspberries, blackberries and wimberries – no need to do anything.

To make the fruit butter, all you need to do is cook the fruit or fruit mixture in a pan until it's soft, adding as little water as you can get away with. Sieve the cooked fruit through a fine mesh, leaving a lovely smooth cream. Weigh the fruit pulp. Then add half the amount of sugar – so if you have 500g pulp, add 250g sugar (you can use any sugar you like – brown sugar gives a richer and more complex flavour than white).

To every 2.2 litres of this mixture, add one-eighth of a teaspoon of salt. This helps bring out the flavours of the fruits. Bring to a rolling boil, stirring constantly, then lower the heat, still stirring, until the mixture starts to thicken. Test as for jam – a blob on a cold saucer should pucker when you push it with your finger. If it's ready, pour into warmed, sterilised jars, and seal.

Spices, too, are good to add to this 'butter'. Blackberries go well with star anise; apples and cinnamon are sensational; and for something completely unexpected, try damsons and lemongrass.

Llanfrynach Hedgerow Summer Pudding

If you want to have a real taste of summer in the winter months – just when you need it – and a reminder of longer (and hopefully warmer) days to come, why not freeze the wild fruits used in this pudding and make it to serve on New Year's Eve? The fresh tanginess of the flavours makes a welcome change from the stodgy fare that's usually associated with this time of year. If you are going to freeze the fruits, make sure you wash and dry them well first.

SERVES 8

8 slices slightly stale white bread, plus extra if needed
Butter, softened
250g elderberries
250g blackberries
375ml water
300g granulated white sugar
750g crab apples (chop and core before freezing)
250g mixed raspberries and wimberries

First prepare your pudding bowl. Cut the crusts from the bread and butter each slice evenly and lightly. Line a transparent glass pudding basin with the bread, placing it butter side out. It's easier to start at the bottom of the bowl and work around the sides. You may make a few mistakes, so have some extra bread to hand. You'll also need to reserve a couple of slices for the top. Set to one side.

Put the elderberries and blackberries into a pan with the water and sugar. Bring to the boil and simmer for 5 minutes. Sieve the berries, retaining the juice, but don't press the fruit since you want it to remain as whole as possible.

Chop and core the crab apples – no need to peel them, they're too small – and cut into 1cm slices. Pop them into the reserved juice and cook in the juice for 10–15 minutes until they are soft. Mash the crab apple, then fold into the cooked fruit and add the raspberries and wimberries. Leave to cool.

Pack the cooled fruit into the pudding basin, making sure there are no gaps, and top with the rest of the bread. Place a plate on top of the bread 'lid', to help press it down, and then place a heavy weight, such as a large stone, on the plate. Excess juice may ooze out over the top of the basin so stand it on a tray rather than have your work surfaces stickily stained with gory red and purple juices. Chill overnight in the fridge.

To turn out the pudding, take the stone off the top. The plate will probably be stuck; don't worry about that. Take your serving dish, put it upside down on the pudding, and then turn the whole lot upside down. Leave for a few minutes, and the pudding should easily detach from the basin, which you can then remove. A transparent basin means that you can see if the pudding has freed itself or not.

Serve with cream and a happy smile.

Acknowledgements

There are many, many people who helped me with all aspects of this book.

First, to my awesome agent, the legendary Isabel Atherton. I'd been working on ideas for this book for some time and Isabel made it happen. Thanks, too, to James Duffett-Smith.

Colossal, huge and enormous thanks to the lovely Rosemary Davidson at Random House, for saying yes. And also – very important, this – for introducing me to the wonderful and beautiful Lizzie Harper, whose illustrations grace the pages of this book. Talk about synchronicity! And to Friederike Huber for her stunning book design and to Simon Rhodes who oversaw production with inimitable calm and flair.

The sharp eyes and red pen of Jan Bowmer proved invaluable; she spotted many glaring errors – and some very subtle ones – and for this I'm profoundly grateful. Thanks also to Sally Sargent, my eagle-eyed proofreader.

George Lister of the fabulous Samphire Restaurant in Whitstable, Kent, and Caroline Dodds very kindly supplied an amazing bottle of sloe gin just when it was needed.

I would like to thank Brychan Llyr at www.tipiwales.co.uk whose love of nature and all things wild and wonderful is a constant source of inspiration.

Thanks also to Jan and Paul at Primrose Organics (www.primroseearthcentre.co.uk), near Hay on Wye, for their kindness and help.

I would like to thank C and S for taking me on a long walk one day and showing me some of the best stretches of hedgerow that I have ever seen. And I'd like to thank Liam for understanding the value of a nice hedge.

Many people very kindly gave me recipes which I have adapted, where necessary, for this book; their details follow opposite.

Recipe
Acknowledgements

CELT (www.celtnet.org)
Spicy Pickled Ash Keys, Bird Cherry Flour, Elderberry and Almond Pie, Haw Syrup, Ribwort Plantain Seed Pudding, Rose Petal Jam

Euroresidentes (www.euroresidentes.com)
Beechnut Turron Nougat

Andi Oliver (at The Moveable Feast – see Facebook)
Chilli Blackberry Syrup (this was adapted by Andi from an original recipe in *Gourmet* Magazine, September 2007)

Gourmet Traveller magazine
Chamomile Panna Cotta

The Splendid Table by Lynne Rossetto Kasper (William Morrow, 1992)
Sweet Chestnut and Ricotta Cheesecake

Liam Fitzpatrick
inspired the recipes for Sticky Sweet Soy Chestnuts, Spicy Caramelised Hazelnuts, Elderflower Sabayon and Meadowsweet Sabayon

Lambeth Band of Solidarity
Cleavers Curry

Pille Petersoo (www.nami-nami.blogspot.com)
Crab Apple Cake, Spicy Crab Apple Marmalade

Cooking Weeds by Vivien Weise (Prospect Books, 2009)
Cleavers Soup, Elderflower Cake, Ground Elder Soup, Dog Daisy Spread, Sorrel Soup

Heathcliffe Bird (www.heathcliffebird.com)
Elderberry Flu Remedy

Stuart Matthewman
inspired the wild garlic flu remedy

Sarah Howcroft (www.shamanism-wales.co.uk)
Hawthorn Fruit Leather, Wild Garlic Bannock Bread, Wild Garlic and Greens Soup

Colin Fox
Elderflower Lemonade

Allie Thomas of Cradoc's Savoury Biscuits – see Facebook)
Fallen Hazelnut and Cheddar Crackers

Robin Harford (www.eatweeds.co.uk)
Himalayan Balsam Seed Curry

Lynne Allbutt (www.lynneallbutt.
com)
Nettle Syrup (and tips on nut
cracking)

Neenah Ellis
Honeysuckle Ice Cream

Eating Flowers by Lucia Stuart
(www.eatingflowers.com)
Elderflower and Rhubarb Ripple Ice
Cream

Prodigal Gardens (www.prodigal-
gardens.info)
Red Clover Lemonade

www.watercress.com
Watercress Hummus, Watercress
Virgin Mary

Bompas and Parr (www.jellymon-
gers.co.uk)
Champagne and Wild Strawberry
Jelly, adapted from their superb book
Jelly with Bompas and Parr (Anova
Books, 2010)

The Traveller's Rest, Talybont-on-
Usk (www.travellersrestinn.com)
Wimberry Trumble

Peter Sommer (www.petersommer.
com)
Samphire Salad

Vegbox (www.vegbox-recipes.com)
Elderberry and Almond Pie

Fi Bird, author of the fabulous *Kids'
Kitchen* (Barefoot Books, 2009) and
proprietrix of www.stirrinstuff.org
Wild Raspberry and Meadowsweet
Jam

Index
of Ailments

gout, 75, 93
growing pains, 90

headache, 90, 124, 161, 177, 195
heart disease, 44
hypertension, 98

immune system, vitamin C, 71, 80,
  98, 133, 164, 173, 181, 199, 204, 210
impatience, 108
indigestion, 117, 148
influenza, 80, 86, 88, 124, 128, 205
insomnia, 195
irritability, 108
irritable bowel syndrome, 148

kidney, infection, 42, 44; stone, 137

laryngitis, 181
laxative, 57, 71, 80, 155, 164, 181
liver, infection, 44, 75; tonic, 57

menstrual, cycle, 90; flow, 103, 191;
  pain, 47, 128, 155, 161
mental tension, 108
migraine, 90

nausea, 47
night sweats, see Fever

respiratory infections, 42, 103,112, 141,
  151, 181, 195. See also Cold,
  Influenza
rheumatism, 24, 42, 75, 90, 117, 124,
  133, 151, 195

scurvy, 121
sedative, 47, 93
skin, infection, 112; complaints, 145,
  161, 199
stomach upset, 16, 128, 145, 187
stress, 151

temperature, see Fever
thirst, 187
throat, sore, 35, 39, 80, 112, 114, 137,
  161, 173, 181
tonsillitis, 181
toothache, 219
toxins, 75

ulcers, 60
urinary tract infection, see Diuretic

wart, 75
whooping cough, 52
wound, 60, 121, 141, 148, 219

Published by Square Peg 2012

4 6 8 10 9 7 5 3

Copyright © Adele Nozedar 2012

Illustrations copyright © Lizzie Harper 2012

'Cynddylan on a Tractor' © by RS Thomas quoted with the kind permission of the Estate of RS Thomas

Adele Nozedar has asserted her right under the Copyright, Designs and Patents Act 1988 to be identified as the author of this work. This book is sold subject to the condition that it shall not, by way of trade or otherwise, be lent, resold, hired out, or otherwise circulated without the publisher's prior consent in any form of binding or cover other than that in which it is published and without a similar condition, including this condition, being imposed on the subsequent purchaser

First published in Great Britain in 2012 by

Square Peg

Random House, 20 Vauxhall Bridge Road,

London SW1V 2SA

www.vintage-books.co.uk

Addresses for companies within The Random House Group Limited can be found at: www.randomhouse.co.uk/offices.htm

The Random House Group Limited Reg. No. 954009

A CIP catalogue record for this book is available from the British Library

ISBN 9780224086714

The Random House Group Limited supports The Forest Stewardship Council (FSC®), the leading international forest certification organisation. Our books carrying the FSC label are printed on FSC® certified paper. FSC is the only forest certification scheme endorsed by the leading environmental organisations, including Greenpeace. Our paper procurement policy can be found at www.randomhouse.co.uk/environment

Typeset and designed by Friederike Huber

Printed and bound in Italy by Graphicom Srl

MIX

Paper from responsible sources

FSC® C013123

FSC www.fsc.org